SpringerBriefs in Applied Sciences and Technology

SpringerBriefs present concise summaries of cutting-edge research and practical applications across a wide spectrum of fields. Featuring compact volumes of 50–125 pages, the series covers a range of content from professional to academic.

Typical publications can be:

- A timely report of state-of-the art methods
- An introduction to or a manual for the application of mathematical or computer techniques
- A bridge between new research results, as published in journal articles
- A snapshot of a hot or emerging topic
- An in-depth case study
- A presentation of core concepts that students must understand in order to make independent contributions

SpringerBriefs are characterized by fast, global electronic dissemination, standard publishing contracts, standardized manuscript preparation and formatting guidelines, and expedited production schedules.

On the one hand, **SpringerBriefs in Applied Sciences and Technology** are devoted to the publication of fundamentals and applications within the different classical engineering disciplines as well as in interdisciplinary fields that recently emerged between these areas. On the other hand, as the boundary separating fundamental research and applied technology is more and more dissolving, this series is particularly open to trans-disciplinary topics between fundamental science and engineering.

Indexed by EI-Compendex, SCOPUS and Springerlink.

More information about this series at http://www.springer.com/series/8884

Dejan Radojčić

Reflections on Power Prediction Modeling of Conventional High-Speed Craft

Dejan Radojčić
Faculty of Mechanical Engineering,
  Department of Naval Architecture
University of Belgrade
Belgrade, Serbia

ISSN 2191-530X   ISSN 2191-5318   (electronic)
SpringerBriefs in Applied Sciences and Technology
ISBN 978-3-319-94898-0   ISBN 978-3-319-94899-7   (eBook)
https://doi.org/10.1007/978-3-319-94899-7

Library of Congress Control Number: 2018951908

This Springer imprint is published by the registered company Springer Nature Switzerland AG
The registered company address is: Gewerbestrasse 11, 6330 Cham, Switzerland

To my teachers:

*Prof. Borivoje Ribar—who taught me the
basics of naval architecture*

*Prof. Borislav Djodjo—who taught me
everything else that was necessary to become
a teacher*

Foreword

This book is about evolution, taking note of the history of technological progress for mathematical models to predict calm water resistance, trim and propeller characteristics for high-performance marine vessels. Dr. Dejan Radojčić has been an activist from early in his academic career to develop mathematical models using emerging techniques with validated predictions using available experimental data.

His goals have focused on calculation procedures to improve naval architect's capability to reliably develop high-performance vessel hydrodynamic designs. With these improved resources, designers are able early on to optimize vessels with respect to their requirements to maximize the performance in their operational environment.

I have been a friend and colleague of Dr. Radojčić for more than 30 years. I have learned much from him and encouraged this documentation of his passion for improving the analytical prediction techniques. This book is one of the many achievements of Dejan's life work resulting in especially useful prediction methods developed with artificial neural networks (ANNs). His recent ANN methods have been used to develop techniques sensitive to hull geometry input resulting in predictions of calm water resistance and trim near to that of towing tank quality.

Now retired after 60 plus years as a designer of high-performance vessels, I believe that you will also find this book to be very useful in your work and career.

Chesapeake, VA, USA
April 2018

Donald L. Blount, P.E.
Founder of Donald L. Blount
and Associates, Inc.

Preface

This work was initially intended to be a review paper, but the manuscript grew—and is now a small book that fits the Springer Brief series. It was actually inspired by and envisaged as an extension of the seminal Blount (1993) paper titled "Reflections on Planing Hull Technology". Hence, the titles are intentionally similar. This work is a summation of the author's insights into, and experiences with, high-speed craft (HSC) design and modeling, lovingly accumulated over a 35-year career in the industry and academia, almost entirely focused on this specific topic.

The essence of the Blount 1993 paper (recently updated to a book in 2014) is today still very relevant. The focus of that work was on the *planing hulls* and is expanded here to the category described as *conventional high-speed craft* that covers the largest number of yachts and boats currently in existence or under construction.

The following statements from the abovementioned work inspired the author to write this book, with the principal statements underlined and with text in italics replacing the original verbiage shown in brackets, as appropriate:

- I have chosen to reflect upon *power prediction modeling of conventional HSC* (planing hull technology) in relationship to my personal experience.
- While reflecting on the past, I scanned many references and was reminded of the extensive quantity of useful data which has been published. Looking at my references, I find that only a limited number are "dog-eared" and most have not been referred to too frequently.
- Over the years, I believe my greatest contribution as a naval architect has been my focus on the *power prediction modeling* (practical applications of technology).

This text focuses on the practical and concrete topics and avoids unnecessary issues, mathematical derivations, and similar rigor. However, it is assumed that the reader has a basic university-level knowledge of ship hydrodynamics. Topics which can be found in other books and in the large number of referenced papers are not repeated; they are regarded as companion sources. The author's personal experiences are reflected in the many routines discussed here, since they were derived by him and his team. It is believed that similar reference source material on high-speed craft hydrodynamics does not exist. Consequently, this book is intended primarily for the naval architects who design and develop various types of conventional high-speed vessels, although it may be of use to anyone who is interested in the design of fast vessels.

The author expects that his colleagues and co-authors from the University of Belgrade, Faculty of Mechanical Engineering, Department of Naval Architecture, will soon disclose some, or all, of the mathematical models and/or programs treated here. The Department's Web site is: www.brodogradnja.org.

Belgrade, Serbia   Dejan Radojčić

Reference

Blount DL (1993) Reflections on planing hull technology. In: 5th power boat symposium, SNAME Southeast Section

Acknowledgements

I would like to thank Donald Blount who unselfishly shared his experience and knowledge over the last three decades. He was always inspirational and ready to support work on new topics. Don's constructive suggestions and reviews of some of my manuscripts, including this one, certainly improved them.

Many thanks to Mike Morabito and Predrag Bojović who gave exceptionally useful comments and suggestions. Aleksandar Simić, Milan Kalajdžić, and Rade Pešterac helped with the diagrams. I am proud that the last four are my ex-students.

The consistent support and patience of my family should also be acknowledged. Normally, this is understood and goes unsaid, but since this book is a summary of my life's work, the well-deserved acknowledgment is appropriate.

About this Book

High-speed craft is very different from conventional ships. This dictates the need, from the very outset, for special treatment in designing high-speed vessels. Professional literature, which is mostly focused on conventional ships, leaves a gap in the documentation of best design practices for high-speed craft.

The various power prediction methods, a principal design objective for high-speed craft of displacement, semi-displacement, and planing type, are addressed. At the core of the power prediction methods are mathematical models, based on experimental data derived on models representing various high-speed hull and propeller series. The regression analysis and artificial neural network (ANN) methods are used as an extraction tool for this kind of mathematical models. A variety of mathematical models of this type are discussed in the book.

The most significant factors for in-service power prediction are bare hull resistance, dynamic trim, and the propeller's open water efficiency. Therefore, mathematical modeling of these factors is a specific focus of the book, although other less significant resistance components and hull-propeller interactions are also addressed. Furthermore, the book includes a summary of most of the power-prediction-relevant literature published in the last 50 years, and as such is intended as a reference overview of best modeling practices.

Note that once these mathematical models have been developed and validated, they can be readily programmed into software tools, thereby enabling the parametric analyses required for the optimization of a high-speed craft design. This book provides the foundational reference for these software tools and their use in the design of high-speed craft. It is aimed at the high-speed craft community in general and particularly at the naval architects who design and develop various types of high-speed vessels.

This book is a summation of the author's insights and experiences accumulated over a 35-year career in the industry and academia, focused almost entirely on high-speed craft design and modeling.

Contents

Abbreviations

AEW	Admiralty Experiment Works, Haslar
ANN	Artificial neural network
ATTC	American Towing Tank Conference
BSRA	British Ship Research Association
CFD	Computational fluid dynamics
CPP	Controllable pitch propeller
DSDS	Delft Systematic Deadrise Series
DTMB (TMB)	David Taylor Model Basin
DTNSRDC (NSRDC)	David Taylor Naval Ship Research and Development Center
DUT	Delft University of Technology
HSC	High-speed craft
HSMV	High-speed marine vessels
IMO	International Maritime Organization
ITTC	International Towing Tank Conference
KCA	Kings College Admiralty (Newcastle)
MARIN (Wageningen)	Maritime Research Institute of the Netherlands
MM	Mathematical model
NPL	National Physical Laboratory
NSS	Naples Systematic Series
NTUA	National Technical University of Athens
PHF	Planing Hull Forms
RINA (INA)	The Royal Institution of Naval Architects
SKLAD	Series tested in the Naval Institute in Zagreb
SNAJ	The Society of Naval Architects of Japan
SNAME	The Society of Naval Architects and Marine Engineers
SPP	Surface piercing propeller
SSPA	Swedish Maritime Research Centre
SVA	Potsdam Model Basin

SWATH	Small waterplane area twin hull
TUNS	Technical University of Nova Scotia
USCG	United States Coast Guard
VTT	Technical Research Centre of Finland
VWS	Versuchsanstalt für Wasserbau und Schiffbau (Berlin)
WEGEMT	EU Marine University Association
WUMTIA (Wolfson Unit)	Wolfson Unit for Marine Technology and Industrial Aerodynamics

Symbols

A_D	Developed propeller blade area (m^2)
A_E	Expanded propeller blade area (m^2)
A_O	Propeller disk area (m^2)
A_O'	Immersed area of SPP (m^2)
A_P	Projected propeller blade area (m^2)
A_P	Projected planing bottom area (m^2)
A_T	Transom area (m^2)
A_X	Maximum section area (m^2)
$A_P/\nabla^{2/3}$	Planing area coefficient
A_T/A_X	Transom area ratio
BAR	Blade area ratio
B_{EF}	Effective planing beam (m)
B_M	Beam at midship ($L_P/2$) (m)
$B_{PA} = A_P/L_P$	Mean beam over chines (m)
B_{PT}	Projected chine beam at transom (m)
B_{PX}	Maximal projected chine beam (m)
$B_{WL} = B = B_X$	Beam of hull on DWL (m)
B_{XDH}	Maximum beam of demihull (catamaran) (m)
$C = 30.1266 \cdot v/(L)^{1/4} \cdot (\Delta/2P_E)^{1/2}$	C factor (note: v in kn; not non-dimensional coef.)
C_A	Correlation allowance
C_{AA}	Air resistance (allowance) coefficient
CA_P	Centroid of A_P forward of transom (m)
$C_B = \nabla/(L \cdot B \cdot T)$	Block coefficient
C_F	Specific frictional resistance coefficient
C_R	Specific residuary resistance coefficient
$C_S = S/(\nabla \cdot L)^{1/2}$	Taylor wetted surface coefficient
$C_{T\nabla} = R_T/(\rho/2 \cdot v^2 \cdot \nabla^{2/3})$	Total resistance coefficient
$C_T^* = T/(\rho/2 \cdot v_{0.7R}^2 \cdot A_O)$	Thrust index (coefficient) of propeller
$C_Q^* = Q/(\rho/2 \cdot v_{0.7R}^2 \cdot A_O \cdot D)$	Torque index (coefficient) of propeller

$C_X = A_X/B \cdot T$	Maximum section area coefficient
$C_\Delta = \nabla/B_{PX}^3$ or ∇/B_X^3	Beam load coefficient
$C_\nabla = C_{DL} = \nabla/(0.1 \cdot L)^3$	Volume displacement coefficient (also in use ∇/L^3)
D	Propeller diameter (m)
$DAR = A_D/A_O$	Developed area ratio
DWL	Designed waterline at rest
$EAR = A_E/A_O$	Expanded area ratio
$Fn_B = C_V = v/(g \cdot B_{PX})^{1/2}$	Beam Froude number
$Fn_h = v/(g \cdot h)^{1/2}$	Depth Froude number
$Fn_L = Fn = v/(g \cdot L_{WL})^{1/2}$	Length Froude number
$Fn_\nabla = v/(g \cdot \nabla^{1/3})^{1/2}$	Volumetric Froude number
$Fn_{\nabla/2}$	Volumetric Froude number of demihull (catamaran)
g	Acceleration of gravity (m/s^2)
h	Water depth (m)
h	Immersion of SPP (m)
h/D	Immersion ratio of SPP
i_e	Half angle of entrance of waterline at bow (deg)
$K_T = T/(\rho \cdot n^2 \cdot D^4)$	Thrust coefficient
$K_T' = T/(\rho \cdot n^2 \cdot D^2 \cdot A_O')$	Revised thrust coefficient for SPP
$K_Q = Q/(\rho \cdot n^2 \cdot D^5)$	Torque coefficient
$K_Q' = Q/(\rho \cdot n^2 \cdot D^3 \cdot A_O')$	Revised torque coefficient for SPP
$L = T_a \cdot \sin(\psi+\tau) + N \cdot \cos(\psi+\tau)$	Vertical propeller force (lift) (kN)
L_C	Chine wetted length (m)
L_K	Keel wetted length (m)
$L_M = (L_K + L_C)/2$	Mean wetted length (m)
L_{OA}	Length overall (m)
L_P	Projected chine length (m)
L_{PP}	Length between perpendiculars (m)
$L_{WL} = L$	Waterline length (m)
L_P/B_{PX} or L_{WL}/B_{WL}	Length beam ratio
$L_P/\nabla^{1/3}$ or $L_{WL}/\nabla^{1/3} = (M)$	Slenderness ratio
$L_P/(\nabla/2)^{1/3}$	Slenderness ratio of demihull (catamaran)
LCB	Longitudinal center of buoyancy (m)
LCG	Longitudinal center of gravity forward of transom (m)
LCG/L_P	Longitudinal center of gravity relative to transom
$\%LCG = (CA_P - LCG) \cdot 100/L_P$	Longitudinal center of gravity aft of A_p centroid (%)
N	Force normal to propeller shaft line (kN)
n or RPM	Propeller rotational speed (1/sec)

$J = v_a/n \cdot D$	Advance coefficient
$J_\Psi = v_a \cdot \cos\Psi/n \cdot D$	Revised advance coefficient for SPP
P	Propeller pitch (m)
P/D	Pitch–diameter ratio
P_B	Brake power (kW)
P_{BTR}	Brake power trial conditions (catamaran) (kW)
P_D	Delivered power (kW)
P_E	Effective power (kW)
P_E^*	Effective in-service power (kW)
p_a	Atmospheric pressure (kPa)
$p_h = \rho \cdot g \cdot h$	Static water pressure (kPa)
p_v	Vapor pressure of water (kPa)
Q	Propeller torque (kNm)
$Q_C = Q/(\rho/2 \cdot D \cdot A_P \cdot v_{0.7R}^2)$	Torque load coefficient
RCG	Rise of center of gravity
R_F	Frictional resistance (kN)
$Rn = v \cdot L/\nu$	Reynolds number
R_R	Residuary resistance (kN)
$R_T = R$	Total bare hull resistance (kN)
R_T^*	Total in-service resistance (kN)
R_{Th}	Total bare hull resistance in shallow water (kN)
$R/\Delta = (R_T/\Delta)_{100000}$	Resistance to weight ratio (for Δ = 100,000 lb = 45.36 t)
R_W	Wave making resistance (kN)
R_{Wh}/R_{Wd}	Shallow water resistance factor
S	Wetted surface area (m^2)
$(S) = S/\nabla^{2/3}$	Wetted surface area coefficient
t	Thrust deduction fraction
$T = T_H$	Hull draught at DWL (m)
T	Propeller thrust (kN)
T_a	Axial propeller force (kN)
$T_h = T_a \cdot \cos(\psi+\tau) - N \cdot \sin(\psi+\tau)$	Horizontal propeller force (kN)
v	Velocity of craft (m/s)
$v_a = v \cdot (1 - w)$	Speed of advance of propeller (m/s)
$v_{0.7R} = [v_a^2 + (0.7\pi nD)^2]^{1/2}$	Resultant water velocity at 0.7R (m/s)
w	Wake fraction
z	Number of propeller blades
$\beta = \arctan \cdot [v_a/(0.7 \cdot \pi \cdot n \cdot D)]$	Hydrodynamic pitch angle at 0.7R
β_{EF}	Effective deadrise angle (deg)
$\beta = \beta_M$	Deadrise angle at midship ($L_P/2$) (deg)
β_{Bpx}	Deadrise angle at B_{PX} (deg)
β_T	Deadrise angle at transom (deg)

γ	Buttock angle (average centerline angle from $L_P/2$ to transom) (deg)
δ_W	Angle of transom wedge (catamaran) (deg)
$\Delta = \nabla \cdot \rho$	Displacement, mass (tons)
$\Delta K_T = K_{Tatm} - K_{Tcav}$	K_T reduction for cavitating conditions
$\Delta K_Q = K_{Qatm} - K_{Qcav}$	K_Q reduction for cavitating conditions
∇	Displacement volume (m^3)
$\varepsilon = \beta_M - \beta_T$	Warp angle (according to DUT terminology twist angle) (deg)
$\varepsilon_B = P_B/\Delta \cdot g \cdot v$	Specific break power (catamaran)
η_B	Propeller efficiency behind the vessel
$\eta_D = P_E/P_D$	Propulsive efficiency (quasi-propulsive efficiency)
$\eta_H = (1-t)/(1-w)$	Hull efficiency
$\eta_O = (K_T/K_Q) \cdot (J/2\pi)$	Propeller open water efficiency
$\eta_P = \eta_H \cdot \eta_R \cdot \eta_S \cdot \eta_O$	Overall (total) propulsive coefficient (OPC)
$\eta_R = \eta_B/\eta_O$	Relative rotative efficiency
η_S	Shaft efficiency (including gearing efficiency)
ν	Kinematic viscosity of water (m^2/s)
ρ	Mass density of water (kg/m^3)
$\sigma = (p_A + p_H - p_V)/(\rho/2 \cdot v_a^2)$	Cavitation number based on advance velocity
$\sigma_{0.7R} = (p_A + p_H - p_V)/(\rho/2 \cdot v_{0.7R}^2)$	Cavitation number based on resultant water velocity at 0.7 radius
τ (θ in some references)	Dynamic (running) trim relative to its value at zero speed (deg)
$\tau_{BL} = \tau_O + \tau$	Baseline trim angle (deg)
$\tau_c = T/(\rho/2 \cdot A_P \cdot v_{0.7R}^2)$	Thrust load coefficient
τ_O	Initial static baseline trim (deg)
Ψ	Shaft inclination relative to buttock (deg)

Chapter 1
Introduction

1.1 Objectives

The main goals of this book are to:

- Review various statistically based Mathematical Models (MM) for power prediction
- Spotlight some very useful MMs
- Encourage the HSC designer to use existing MMs.

The reason for the abovementioned objectives is the author's belief that a large number of recently published papers are too complex to be useful in everyday practice. As a consequence, practicing naval architects in need of a power prediction, are indirectly forced to rely on commercial software whose essence is often not properly understood. Moreover, the few decades-old experimental results, MMs, etc., are often considered to be archaic, particularly by the younger engineers, and are hence frequently marginalized, although they have not been replaced by better MMs or routines.

Moreover, MM development is an evolutionary process, i.e. MM developers should be familiar with what their predecessors have done. Thus, one of the objectives of this work is to aid new MM developers by reviewing the existing models along with their principal characteristics and tradeoffs.

The core of this work are statistically based MMs based on the results of model experiments of various HSC series. A variety of regression analysis and lately Artificial Neural Network (ANN) methods were used to develop these MMs. The

---

The original version of this chapter was revised: For detailed information please see correction. The correction to this chapter is available at https://doi.org/10.1007/978-3-319-94899-7_8

D. Radojčić, *Reflections on Power Prediction Modeling of Conventional High-Speed Craft*, SpringerBriefs in Applied Sciences and Technology, https://doi.org/10.1007/978-3-319-94899-7_1

resulting MMs can be easily programmed into software tools, thereby enabling the parametric analyses required for design optimization.

1.2   Conventional High-Speed Craft (HSC)[1]

The term *conventional HSC* is applied here to the high-speed craft of displacement, semi-displacement, and planing types which achieve speeds that include and/or exceed the main resistance hump. This corresponds roughly to the length Froude number $Fn_L > 0.4$, volume Froude number $Fn_\nabla > 1$, and beam Froude number $Fn_B > 0.5$, approximately resulting in the following classifications:

- IMO according to which "HSC is a craft capable of maximum speed equal to or exceeding $3.7 \cdot \nabla^{0.1667}$ m/s" (which is actually equivalent to $Fn_\nabla > 1.18$), and
- ITTC according to which "high-speed marine vehicles are defined to be vessels with a design speed corresponding to a Froude number above 0.45, and/or a speed above $3.7 \cdot \nabla^{0.1667}$ m/s, and/or where high trim angles are expected, or for dynamically supported vessels".

The lowest speed for the dry or fully-vented transom approximately corresponds to $Fn_L > 0.3$ or $Fn_\nabla > 1$ (Blount 2014).

The relatively wide speed range of HSC should be emphasized. Namely, a single vessel may travel in displacement ($Fn_L < 0.40$), semi-displacement ($0.40 < Fn_L < 0.65$), semi-planing[2] ($Fn_L > 0.65$ but $Fn_\nabla < 3.0$), and pure planing regimes ($Fn_\nabla > 3.0$). Note that the approximate Fn values given in parenthesis are typical for slenderness ratio $L/\nabla^{1/3} \approx 6.0$; see Blount (1995, 2014). Each regime has its peculiarities, so that different parameters are necessary to model the performance for each of them. For instance, for speeds below $Fn_L \approx 1$ it is better to use Fn_L than Fn_∇ and vice versa (see above). This makes modeling relatively difficult, as it is desirable to describe the performance over the entire sailing regime with the same parameters (input variables) and if possible, with a single continuous equation.

1.3   Resistance, Propulsion, and Power Prediction

Typically, one of the main design objectives is the minimization of power. To achieve this, optimization of the whole system—including resistance, propulsion, and engine

---

[1] According to the 16th ITTC HSMV Panel, HSC are grouped and divided into following types: (a) Hydrofoils, (b) Hard chine planing craft, (c) Round bilge semi-planing, (d) SWATH ships, (e) Air Cushion vehicles (amphibious), (f) Surface Effect Ships (non-amphibious), and (g) Others. HSC belonging to groups (b) and (c) are treated here.

[2] According to Savitsky (2014) "Vessels operating in the speed range between hull-speed and planing inception speed are totally supported by buoyant forces, hence the use of the terms 'Semi-Planing' or 'Semi-Displacement' hulls is inappropriate". Nevertheless, the author of present book decided to continue using those well-established terms, for reasons of continuity and tradition.

(with the gearbox)—is necessary, since separate optimization of the components, in isolation to the rest of the system, does not necessarily result in the same answer. The holistic system optimization, i.e. *complex* or *integrated approach*, is much more important for the HSC than for conventional displacement vessels. However, in order to achieve some clarity, the current work presents the subject in the conventional way, i.e. MMs for resistance and propulsion predictions are presented separately, while the integrated approach is elaborated in Chap. 6. Separate treatment of resistance and propulsion enables independent investigation of the influence of hull and of propeller parameters on hydrodynamic performance. With an integrated approach these influences have to be examined simultaneously, which significantly complicates the evaluation.

One of the key lessons learned is that power prediction for a desired speed (or vice versa, speed prediction for installed power) must be considered from the very early design phases. Moreover, initial predictions of speed should be rechecked whenever any design changes or modifications that affect performance are initiated. Thus, neglecting power prediction during the various design phases often results in degraded performance, and the actual achieved speed is almost certainly below the predicted speed.

For power prediction (P_B) evaluation of in-service total resistance (R_T*) and overall propulsive efficiency (η_P) are necessary as

$$P_B = P_E*/\eta_P = R_T* \cdot v/\eta_P$$

where P_B and P_E are effective and brake powers respectively. Note that routines that model bare hull resistance are usually valid for calm and deep water (denoted here as R_T). Various additional factors should be taken into account to predict actual in-service performance (e.g. appendages, air resistance, waves, restricted waterways, etc.) and "*" is used after P_E and R_T to denote this. Detailed subdivision of HSC total resistance, as well as various components that form R_T* are given, for instance, in Müller-Graf (1997a, b), and will be discussed later.

The most significant portion of the overall propulsive efficiency (η_P) is the open water efficiency (η_O) of the selected, presumably optimal, propeller. Assessment of a propeller's open water efficiency, however, is not a straight forward procedure; it is actually a separate task to determine the best (i.e. optimal) propeller for a given set of requirements. The optimal propeller should produce thrust that overwhelms in-service resistance for both design and off-design conditions. Consequently, the open water efficiency of a propeller is a result of a complete propulsion analysis.

As is well known, the dynamic trim angle (τ) is very important for HSC resistance, propulsion, and performance in general. In a way, hull resistance mirrors the dynamic trim angle, and hence dynamic trim angle evaluations usually go hand-in-hand with the resistance evaluations. The relationship between hull resistance and dynamic trim is particularly pronounced for the hump speeds, which HSC cannot avoid, so that modeling this range is of the utmost importance.

Thus, the most significant factors that must be reliably evaluated for power prediction are:

- Bare hull resistance (R_T) and Dynamic trim (τ), and
- Propeller's open water efficiency (η_O).

Hence, this work is focussed on the mathematical modeling of these factors.

Other quantities in the abovementioned equation are also important, but are typically less significant. They are essential parts of power prediction routines, but are not the subject of this work per se. For the sake of completeness, these components are briefly discussed in Chap. 5.

1.4 Common Mistakes

The most common power prediction mistakes are:

- An incorrect prediction model (MM) is selected, i.e. the vessel under analysis has different characteristics than those upon which the MM is based.
- Violation of the boundaries of applicability of MM.

Therefore, it is important to know how a particular MM was developed, the constraints and assumptions it used in formulation. This information is often missing, particularly when commercial power prediction software tools are used. For example, several very important hull or propeller characteristics may be "masked" (i.e. MMs are inherently valid for a particular hull form or propulsor type). These hidden characteristics, which may not be required by a MM should be regarded as additional and prescribed quantities that limit the applicability of a given MM. In other words, wise usage of readymade computer programs requires that the designers are familiar with the characteristics of the hull- and propeller-series the MM is based on, as well as with the technique used for its derivation.

MacPherson (1993) gives a good review of dos and don'ts concerning numerical prediction techniques. Some recommendations deserve to be cited:

- Not all methods are appropriate for all problems.
- Know your prediction method. The numerical procedure must be fully understood.
- Complete and reliable program cannot ignore the user. The human interface is very important.
- Numerical methods cannot eliminate model testing.

1.5 Excluded Topics

1.5.1 Resistance Evaluation Using Empirical Methods

Not addressed in this work is the Savitsky (1964) method. It is based on equations for prismatic hull forms and is by far the most frequently used amongst various empirical methods. The other planing hull resistance prediction methods are mentioned in Almeter (1993). Savitsky's method is applicable for higher planing speeds where hydrodynamic forces are dominant. Note that Savitsky's method was modified a few times (see for instance Blount and Fox 1976; Savitsky 2012).

1.5.2 Resistance Evaluation Using Computational Fluid Dynamics (CFD)

It is believed that the CFD-based methods will become common everyday tools in the future. At this time however, CFD still depends very much on interpretation of the simulated results by the user. Therefore, these methods are typically not yet sufficiently mature to be used by regular engineers in everyday engineering practice. CFD's subjective nature (Molland et al. 2011; Almeter 2008) is also not yet practical for application within broader numerical optimization tools (typically nonlinear multi-criterion optimization with constraints), where evaluations of resistance and propeller efficiency are just segments of an integrated approach. CFD applied to HSC is given for instance in Savander et al. (2010), Brizzolara and Villa (2010), Garo et al. (2012), and De Luca et al. (2016).

Actually, CFD and the MMs treated here are complementary methods, although they are fundamentally different techniques. Namely, MMs based on experimental data provide a low-cost and reasonably accurate preliminary design tool. If needed, further analysis, improvements and hull/propeller adjustments should be done using CFD and/or tow tank tests. Furthermore, designers typically do not do their own CFD evaluations; these are usually done by the CFD specialists, thereby resulting in 'once-removed' relationships that are similar to experimental facilities (tow tank, cavitation tunnel etc.). This may change in the future as CFD tools become more practical and useful to designers.

The state-of-the-art viewpoint on statistical power performance predictions and CFD, is given by Van Hees (2017):

- Statistical methods are fast, while CFD (including 3D hull modeling etc.) require time.
- In practice, CFD is used to supplement statistical methods, with the objective to further optimize the hull form.

1.5.3 Other Excluded Topics

High-Speed Ships

MMs based on the experimental series with hull forms that resemble displacement ships more than HSC are not treated here, although they are valid for relatively high speeds (Fn_L approaching 1 or so). In other words MMs for high-speed ships are not the subject of this work (for instance, Fung and Leibman 1993; Bojović 1997, etc.).

Not Released Mathematical Model

MMs for hull/propeller series for which complete and usable MMs have not been released are excluded. For instance, the following high-speed hull and propeller series are valuable, but have not been addressed here:

- MARIN systematic series of fast displacement hulls consisting of no less than 33 models, see Kapsenberg (2012), or
- SVA high-speed, 3-bladed, inclined shaft propeller series consisting of 12 models; see Heinke et al. (2009).

Commercial Software

MMs found in commercial software programs are also excluded. Note, however, that most of them are based on the routines that are discussed here.

Waterjets

Waterjets in general, as for their sizing cooperation with the waterjet manufacturer is usually required.

References

Almeter JM (1993) Resistance prediction of planing hulls: state of the art. Mar Technol 30(4)
Almeter JM (2008) Avoiding common errors in high-speed craft powering predictions. In: 6th International conference on high performance marine vehicles, Naples
Blount DL (1995) Factors influencing the selection of hard chine or round bilge hull for high Froude numbers. In: Proceedings of the 3rd International conference on fast sea transportation (FAST '95), Lubeck-Travemunde
Blount DL (2014) Performance by design. ISBN 0-978-9890837-1-3
Blount DL, Fox DL (1976) Small craft power prediction. Mar Technol 13(1)
Bojović P (1997) Resistance of AMECRC Systematic series of high-speed displacement hull forms. In: High speed marine vehicles conference (HSMV 1997), Sorrento
Brizzolara S, Villa D (2010) CFD simulation of planing hulls. In: 7th international conference on high performance marine vehicles, Melbourne Florida
De Luca F, Mancini S, Miranda S, Pensa C (2016) An extended verification and validation study of CFD simulations for planing craft. J Ship Res 60(2)
Fung SC, Leibman L (1993) Statistically-based speed-dependent powering predictions for high-speed transom stern hull forms. Chesapeake Section of SNAME
Garo R, Datla R, Imas L (2012) Numerical simulation of planing hull hydrodynamics. In: SNAME's 3rd Chesapeake power boat symposium, Annapolis
Heinke HJ, Schulze R, Steinwand M (2009) SVA high speed propeller series. In: Proceedings of 10th International conference on fast sea transportation (FAST 2009), Athens
Kapsenberg G (2012) The MARIN systematic series fast displacement hulls. In: 22nd International HISWA symposium on yacht design and yacht construction, Amsterdam
MacPherson DM (1993) Reliable Performance Prediction Techniques Using a Personal Computer. Mar Technol

Molland AF, Turnock SR, Hudson DA (2011) Ship resistance and propulsion—practical estimation of ship propulsive power. Cambridge University Press, ISBN 978-0-521-76052-2

Müller-Graf B (1997a) Part I: Rresistance components of high speed small craft. In: 25th WEGEMT school, small craft technology, NTUA, Athens—ISBN Number: I 900 453 053

Müller-Graf B (1997b) Part II: Powering performance prediction of high speed small craft. In: 25th WEGEMT school, small craft technology, NTUA, Athens—ISBN Number: I 900 453 053

Savander BR, Maki KJ, Land J (2010) The effects of deadrise variation on steady planing hull performance. In: SNAME's 2nd Chesapeake power boat symposium, Annapolis

Savitsky D (1964) Hydrodynamic design of planing hulls. Mar Technol 1(1)

Savitsky D (2012) The Effect of bottom warp on the performance of planing hulls. In: SNAME's 3rd Chesapeake power boat symposium, Annapolis

Savitsky D (2014) Semi-displacement hulls—a misnomer? In: SNAME's 4th Chesapeake power boat symposium, Annapolis

Van Hees MT (2017) Statistical and theoretical prediction methods. Encycl Marit Offshore Eng. (Wiley)

Chapter 2
Mathematical Modeling

2.1 Statistical Modeling

Mathematical modeling which is of interest for present work belongs to the *predictive modeling* class, as opposed to *explanatory* or *descriptive* modeling. According to Shmueli (2010): "Predictive modeling is a process of applying a statistical model or data mining algorithm to data for the purpose of predicting new or future observations… The goal is to predict the output value (Y) for new observations given their input values (X)". The modeling process segregated into a set of steps is described in Fig. 2.1 (Shmueli 2010).

Note that the first three steps are usually performed by one team and the rest by another, although it would be desirable if the entire process was executed by a single multidisciplinary team. In addition, the entire modeling process (steps 1–8) is usually performed solely by the engineers, although knowledge of subject-matter mathematics, specifically statistics, is desirable (see for instance Weisberg 1980; Draper and Smith 1981).

The basic steps given in Fig. 2.1 are clear and logical, and MM developers naturally follow them. Note that Step 4, abbreviated EDA (Exploratory Data Analysis), usually requires transformation of the available data into a format suitable for mathematical modeling. EDA is a very important step because various variables, and their eventual transformations, should be considered at this point in the process. The choice of dependent (target), and most influential independent (input) variables (Step 5); statistical data modeling tool, i.e. MM extraction methods (Step 6); evaluation and selection of final MM amongst several considered (Step 7); and use of recommended MM together with reporting (Step 8); all follow Step 4 and are therefore affected by the decisions made there.

© The Author(s), under exclusive licence to Springer Nature Switzerland AG 2019
D. Radojčić, *Reflections on Power Prediction Modeling of Conventional High-Speed Craft*, SpringerBriefs in Applied Sciences and Technology,
https://doi.org/10.1007/978-3-319-94899-7_2

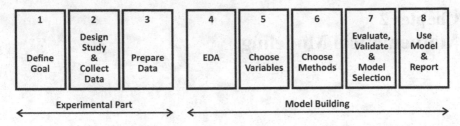

Fig. 2.1 Basic steps in statistical modeling

Datasets as treated here are usually scarce, so that data partitioning[1] is often avoided and all available data is used for building of the MM. This, however, requires thorough MM validation (Step 7), particularly stability checking (possibility of waving between the tested values). Moreover, the use of the entire dataset produces repeatable results, as the holdout samples are usually randomly selected.

The reliability of a MM (*predictive accuracy* or *predictive power* according to Shmueli 2010) is very important, particularly when MM is applied to everyday design problems when the correct value is not known, and the user relies on the MM's predictive accuracy. Therefore, during the development phase, the following should be checked in order to verify the derived model:

- Statistics of the accuracy of the model versus the data set used to develop the model
- Discrepancies between evaluated and measured values
- Behaviour of the model between the data points, where there are no measurements (naturally within the applicability boundaries).

2.1.1 Statistical Modeling Applied to Ship Data

The author of this work never considered parameters of no physical significance or no physical meaning to be primary input variables, despite their possible high statistical correlation. Correlation analysis among the independent variables and versus the dependent variable, was always performed in order to ensure the validity of the selection. When methodical series data is used, the input variables are usually the same as the parameters varied during the model-based series' testing.

Note that the main disadvantage of the statistical data modeling tools is that only a limited number of variables are used to adequately describe the HSC's hull form and loading over a relatively wide speed range. For this reason, the secondary hull

---

[1]Data partitioning, to data that are used for MM development (often called the training set), and to the holdout sample which MM "did not see" (expression from Shmueli 2010) is not unusual. The purpose of the holdout sample is to validate the MM, thus it is also named the validation set.

form parameters are important and should be regarded as a kind of supplement to the MM.

To clarify, the input parameters for a HSC could for example be $L/\nabla^{1/3}$, L/B, and B/T. These hull and loading parameters, although most significant, do not reflect the hull form, i.e. whether it is a hard chine or a round bilge type, and if hard chine then whether it is wide- or narrow-transom, etc. Additional hull description is obviously necessary. This is provided through the secondary hull parameters. However, the secondary hull parameters are often not explicitly specified, but the MM is instead described as being valid for, for instance, the NPL series. This therefore means that the MM is based on the semi-displacement round bilge hulls whose secondary parameters are $C_B = 0.397$, $LCB = 6.4\%L$ aft. amidship, $A_T/A_X = 0.52$, etc. (see Table 3.1). Thus, the additional information typically given in the form of a comment such as "MM is valid for (or is based on) the NPL series" is a very important supplement of the MM. Similarly, the secondary parameters for the propellers would include information about the blade shape and section etc. Naturally, the MM user must be aware of this fact.

Statistical power performance predictions for conventional ships, with a focus on the developmental philosophy of prediction methods, are discussed in a related paper (Van Hees 2017). Amongst the observations is that the statistical methods should be "refreshed" when some new data is available.

2.2 Model Extraction Tools

The author used two methods—statistical data modeling tools—to extract (i.e. develop) the mathematical models for prediction of resistance and propulsive coefficients:

- Regression analysis, and
- Artificial Neural Networks (ANN).

Note that there are several types of regression and ANN methods, but further elaboration on this topic is beyond the scope of the present text.

2.2.1 Regression Analysis

With regression analysis, and in particular with the linear multiple regression analysis (e.g. Weisberg 1980), the independent variables consist of two sets of input data:

- Basic independent variables, and
- Various transformations of basic independent variables.

For the simplest case, with only two basic independent variables x_1 and x_2 (e.g. assuming $x_1 = L/\nabla^{1/3}$, $x_2 = Fn_\nabla$ and dependent variable $Y = R/\Delta$), the additional transformed variables could then be $x_3 = x_1 x_2$, $x_4 = x_1^2$, ... Thus the number of terms in the initial polynomial equation rapidly increases i.e.

$$Y = a_0 + a_1 x_1 + a_2 x_2 + a_3 x_3 + a_4 x_4 + \cdots + a_n x_n$$

where a_0, $a_1, \ldots a_n$ are the coefficients determined by the regression analysis. Note that this is a linear equation, although the basic variables can be transformed for the purpose of simulating nonlinear relationships.

Many different transformations of both independent and dependent variables have been tested by the author, forming equations of, for instance, logarithmic, exponential, or reciprocal types that produced different curves. However, it has been found that when the number of equation terms is relatively high, and the cross-products and different powers of independent variables are used, then there are no great differences in the results. That is, the original polynomial form is sufficient.

Therefore, despite of the fact that the number of basic parameters is typically five or less, the initial polynomial equation, produced by standard automated polynomial fitting, may have 100 or more terms. Consequently, if a certain hull characteristic is not represented directly through the basic hull parameters, then it will most likely be represented indirectly, through one of the many polynomial terms that appear in that initial equation.

By applying a step-by-step procedure and statistical analysis, a subset of significant terms is chosen and less significant variables are eliminated, resulting in the final equation. This final equation, often called "the best equation", comprises considerably fewer independent variables than the initial polynomial equation. The weakness of this approach is that it assumes that there is a single optimal subset of terms in the equation; whereas in fact, usually there may be several sets of terms that work equally well.

In general, more terms in the equation enable better fitting to the data that the MM is based upon, but the interpolated results (those between the data points) may be poorer. That is, a larger number of terms results in more waving in the function, i.e. a less-smooth curve or surface.

The author's experience spans the period from early custom-made programs written in BASIC through to the commercial PC software now in common use, with the regression routines evolving from the so called *backward elimination*, to *forward selection*, and to the more sophisticated *multiple stepwise methods* (combination of previous). This experience indicates that successful application of any of the approaches requires the user to have sufficient subject matter knowledge to choose the best subset (i.e. to obtain the "best equation"). Implementation of the above-mentioned process actually does require an understanding of statistics and statistical methods.

2.2.2 Artificial Neural Network (ANN)

The Artificial Neural Network is a nonlinear statistical data modeling technique that can be used to determine the complex relationships between dependent and independent variables. A brief discussion of the potentials of ANN techniques, and probably some of the first applications of the novel ANN tools in marine design and modeling, are given in Mesbahi and Atlar (2000), Mesbahi and Bertram (2000), Koushan (2001), etc. Amongst the conclusions drawn was that ANN can be successfully used as an alternative to regression analysis.

There is a family of ANN methods that may be used for the derivation of the MMs. The author and his team used a software tool aNETka 2.0 (see Zurek 2007) based on a feed-forward type of ANN routine with a back-propagation algorithm; see Rojas (1996) for instance.

With ANN more attention is paid to the selection of independent variables than with regression analysis (see Radojčić et al. 2014). Specifically, the independent variables must be carefully chosen at the very onset of the fitting process, because the final model is based on the selected input parameters (which form the input layers for ANN). Use of incorrect, or insufficient, independent variables may result in an erroneous MM, with for example, the dependent variable being insensitive to the variations in a poorly-chosen set of input variables. On the other hand, if too many independent variables are assumed, validation of the model stability becomes considerably more complex.

With ANN, in order to find a 'good' solution, in terms of accuracy, reliability, or applicability, the number of layers and number of neurons in each layer must be selected by the user (MM model developer) in advance; see Fig. 2.2. The number of layers and type of activation function are usually constrained by the software used.

Activation or transfer function convert the input signal of a node to an output signal, which then becomes the input signal for another node in the next layer. The nonlinear activation function enables nonlinear transformation between input and output. It also facilitates neural network learning. Without it, the neural network would be a kind of linear regression model. Various activation functions, including linear, sigmoid, and hyperbolic tangent functions were tested, and the sigmoid

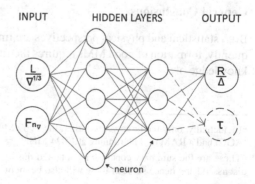

Fig. 2.2 ANN with 2-5-3-1 structure for single-output, or 2-5-3-2 structure (dashed lines) for multiple-output. Input variables are $L/\nabla^{1/3}$ and Fn_∇, while output variables are R/Δ, or R/Δ and τ, for single- and multiple-output, respectively

function (sig $= \frac{1}{1+e^{-x}}$, S-shaped curve) was found to produce the best results and was hence adopted by the author.

An illustration of ANN application to catamaran resistance is given in two related papers by the same authors (Couser et al. 2004; Mason et al. 2005). Different ANN network architecture was investigated, i.e. number of layers and number of neurons were systematically varied. ANN was implemented directly in treatment of the test data, so that multidimensional data smoothing was obtained as a side effect.

ANN technique with multiple outputs (see Fig. 2.2), as opposed to a single output, were applied lately for generating MMs for R/Δ and τ, with very encouraging results (see Radojčić et al. 2017; Radojčić and Kalajdžić 2018).

2.3 Hardware

Statistically based power prediction methods emerged with the proliferation of the use of computers in everyday engineering practice. In the early days of MM development computer power and its peripherals were an important factor, primarily because in practice they constrained the complexity of the MMs. For example, the first MMs developed by the author in the 1980s, were done using successively, SHARP MZ80K (RAM 32 kb, with an audio-tape as external memory), Apple II + (RAM 48 kb, floppy drive), and Olivetti M21 (RAM 640 kb, hard disk). The MMs produced were relatively simple compared to the modern models, and those hardware platforms would be more than insufficient nowadays.[2] Note however that despite the fact that the computational capability has subsequently evolved exponentially, model development is still a lengthy process since the complexity of MMs has also grown. Thus, the selection of the "best equation" amongst many good options, stability checking, validation etc. is still a complex and a time-consuming job, regardless of the virtually unlimited computational power available today.

2.4 Conclusions on Mathematical Modeling[3]

General Conclusions

Both statistical and physical perspectives are important for correct modeling. Consequently, formation of good MMs requires the interdisciplinary approach, i.e. specific knowledge of:

---

[2]Incidentally, only ten or so years earlier, the 1969 Moon-landing Apollo 11 Guidance Computer (AGC) had a RAM of 2 kb running at 1.024 MHz!

[3]These are the summary conclusions obtained during the derivation of the mathematical models discussed here, hence some of these will also be mentioned later.

- Physics of HSC hydrodynamics.
- Statistics, curve fitting, regression analysis, ANN technique etc.
- Procedures for developing and checking MM (choosing "the best equation" out of the many derived candidates).
- Optimization techniques (i.e. what type of MM can be used in a larger numerical optimization tool—type of numerical expression suitable for computer evaluation, numerical boundaries of applicability etc.).

Conclusions on Application of Regression Analysis and ANN

- For extraction of MMs, ANN requires less of the user's manual interference than regression-based methods.
- Regression analysis seems to be more convenient than ANN for modeling simpler relationships with the lesser number of input variables and vice versa.
- Regression analysis (stepwise method) allows screening and rejection of less significant polynomial terms, which is not the case with ANN, where the number of terms is defined at the very beginning.
- Regression based MMs were stiff, at least the MMs of polynomial form. ANN based MMs, with complex relationship between dependent and independent variables and with many equation terms, proved to be more elastic. Moreover, the interpolated data (those between the data points) did not show instability.
- ANN with a larger number of hidden layers usually produced MMs which could better fit the data, but in return those MMs were usually not stable between the data points.
- MMs developed by ANN, as opposed to those developed by regression analysis, allowed limited extrapolation,[4] although as a rule bounds of applicability should not be violated. This is due to the complexity of MMs derived by ANN which may not show an abrupt change of character beyond the applicability zones.
- Once the MM is developed and validated, no further knowledge of ANN or regression technique is needed. Although this sounds logical, the author's experience is that potential MM users tend to be hesitant whenever ANN is mentioned, probably because ANN, in general, has broad applicability and requires specific expertise.

References

Couser P, Mason A, Mason G, Smith CR, von Konsky BR (2004) Artificial neural network for hull resistance prediction. In: 3rd international conference on computer and IT applications in the maritime industries (COMPIT '04), Siguenza

Draper N, Smith H (1981) Applied regression analysis, 2nd ed. Willey

Koushan K (2001) Empirical prediction of ship resistance and wetted surface area using artificial neural networks. In: Cui Wei-Cheng, Zhou Guo-Jun, Wu You-Sheng (eds) Practical design of ships and other floating structures. Elsevier Science Ltd, Amsterdam

---

[4]Similar conclusion was derived in Couser et al. (2004).

Mason A, Couser P, Mason G, Smith CR, von Konsky BR (2005) Optimisation of vessel resistance using genetic algorithms and artificial neural networks. In: 4th International conference on computer and IT applications in the maritime industries (COMPIT '05), Hamburg

Mesbahi E, Atlar M (2000) Artificial neural networks: applications in marine design and modelling. In: 1st International conference on computer and IT applications in the maritime industries (COMPIT '2000), Potsdam

Mesbahi E, Bertram V (2000) Empirical design formulae using artificial neural nets. In: 1st International conference on computer and IT applications in the maritime industries (COMPIT '2000), Potsdam

Radojčić D, Kalajdžić M (2018) Resistance and trim modeling of Naples hard chine systematic series. RINA Trans Int J Small Craft Technol. https://doi.org/10.3940/rina.ijsct.2018.b1.211

Radojčić D, Zgradić A, Kalajdžić M, Simić A (2014) Resistance prediction for hard chine hulls in the pre-planing regime. Polish Marit Res 21(2):82. (Gdansk)

Radojčić DV, Kalajdžić MD, Zgradić AB, Simić AP (2017) Resistance and trim modeling of systematic planing hull Series 62 (With 12.5, 25 and 30 Degrees Deadrise Angles) using artificial neural networks, Part 2: mathematical models. J Ship Prod Des 33(4)

Rojas R (1996) Neural networks—a systematic introduction. Springer, Berlin

Shmueli G (2010) To explain or to predict. Stat Sci 25(3)

Van Hees MT (2017) Statistical and theoretical prediction methods. Encycl Mart Offshore Eng. (Wiley)

Weisberg S (1980) Applied linear regression. Wiley, New York

Zurek S (2007) LabVIEW as a tool for measurements, batch data manipulations and artificial neural network predictions. In: National Instruments, Curriculum Paper Contest, Przeglad Elektrotechniczny, Nr 4/2007

Chapter 3
Resistance and Dynamic Trim Predictions

3.1 An Overview of Early Resistance Prediction Mathematical Models

The first application of regression analysis (actually of the 200 year old Gauss least square method) for prediction of ship resistance is believed to have been used for the design of trawlers (few papers were published by a single author, e.g. Doust 1960). Model-test data of residuary resistance were curves fitted for various speeds; six hull form and loading parameters important for the trawlers were chosen as the independent variables, but an estimation of their relative significance was not reported. In that work, the number of equation terms for resistance prediction was relatively large—typically 30 and in some cases even 86—depending on the speed-length ratio. Accuracy, reported as the differences between measured and calculated values, was acceptable at around 3% for 95% of cases, though it was lower for the hump speeds, where residuary resistance fluctuated more.

Sabit performed separate regression analyses of resistance data for various merchant ships series—BSRA, 60, SSPA (for instance Sabit 1971), and special attention was paid to the correlation amongst the polynomial terms. Evaluated regression coefficients consisted of only up to 16 terms for each Froude number. From the statistical viewpoint, Sabit's approach is considered as more advanced than Doust's.

Van Oortmerssen (1971) produced a single equation for a range of speeds, by using Havelock wave-making resistance theory from 1909. Despite of the fact that it included about 50 polynomial terms, this single equation MM, applicable for trawlers and tugs, is less accurate than Doust's. A similar approach, but for different vessel types, was later followed by other authors from MARIN including a well-known Holtrop and Mennen method (Holtrop and Mennen 1982).

The original version of this chapter was revised: For detailed information please see correction. The correction to this chapter is available at https://doi.org/10.1007/978-3-319-94899-7_8

© The Author(s), under exclusive licence to Springer Nature Switzerland AG 2019
D. Radojčić, *Reflections on Power Prediction Modeling of Conventional High-Speed Craft*, SpringerBriefs in Applied Sciences and Technology,
https://doi.org/10.1007/978-3-319-94899-7_3

Farlie-Clarke (1975) focused on the use of statistical methods in interpretation and evaluation of ship data. Linear and nonlinear least-square methods were applied, resulting in much smaller MMs, with fewer than 10 terms. The author of the present work has adopted the Farlie-Clarke (1975) approach, and has leveraged it in development of multiple MMs throughout his career.

Fung published a series of papers in the 1990s (e.g. Fung 1991; Fung and Leibman 1993), reporting on the application of multiple linear regression analyses to very large databases of transom-stern ships (Fn_L up to 0.9), covering in some cases even 700 ships (10,000 data points). An extensive overview of the application of statistically based regression analysis to ship performance data, and of statistics as a modeling tool, is given in Fung (1991). In the abovementioned references the in-depth analyses and validations of resistance predictions, as well as the limitations of the use of statistically based MMs for resistance predictions are given, hence this may be regarded as a turning point in the MM derivation practices used in naval architecture.

3.2 Types of Mathematical Models for Resistance Prediction

3.2.1 Random Hull Forms Versus Systematical Hull Forms

From the previous overview it follows that regression analysis has been successfully used to analyze the resistance data for random hull forms (e.g. Holtrop and Mennen 1982; Fung 1991; Fung and Leibman 1993) and methodical series (Sabit 1971).

If several random hull forms are considered, then the individual characteristics of each hull or series (i.e. the secondary hull form parameters) cannot be taken into consideration. Namely, the secondary hull form parameters may be lost among the primary parameters, even if many independent (explanatory) variables are introduced into the MM. Therefore, it is often better to narrow down the applicability of the MM to a specific methodical series, and thus to increase the reliability of the model. With this approach, only the primary hull form parameters are modeled explicitly, and it is up to the user to consider that the subject hull must correspond to the series' hull form, and hence to factor in the characteristics which are not explicitly encompassed. The disadvantage of this approach however, is that multiple MMs are required, each for a given methodical series or a group of similar hull forms, instead of having a single MM needed for random hull form approach.

The author of the present work has used both approaches. In the first case (random hull forms), special attention was paid to formation of a database, i.e. a database consisting of multiple but similar hull forms was assembled (hence was not so random), so that the secondary hull form parameters, assumed to be similar, were also factored in. In general, with this approach, it is very important to choose representative hull parameters for the whole database (see Radojčić et al. 2014a). Concerning planing hulls for instance, choosing *effective beam* and *effective deadrise* is of extreme importance; see Blount and Fox (1976) and Savitsky (2012).

Depending on the specific problem, in some cases it may be more convenient to rely upon a MM based on random hulls, in other cases on a methodical series, or often on both.

3.2.2 Speed-Independent Versus Speed-Dependent

There are two general types of equations for resistance evaluation that have evolved over time (see Fung 1991), each having some advantages and disadvantages:

- Speed-independent models (e.g. Sabit 1971; Fung 1991) where separate equations are generated for each discrete speed, since the speed is not included as an independent variable.
- Speed-dependent models (e.g. van Oortmerssen 1971; Fung and Leibman 1993) with vessel's speed included as an independent variable.

The advocates of speed-dependent models claim that the predicted resistance in speed-independent models often does not vary properly with speed, since the resistance computed at one speed is not directly linked to that at another speed. This is because the speed variable is not explicitly included in the regression with this approach. The accuracy of the speed-independent models, however, is believed to be somewhat better at given characterized speeds, since independent equations are developed at each speed point.

Speed-dependent MMs can be further segregated into those which are based on some wave-making theory (as in van Oortmerssen 1971; Fung and Leibman 1993 for instance), and those which are not based on the wave-making theory. MMs which are based on a wave-making theory should realistically represent the position of humps and hollows (which are probably only important at lower speeds). These MMs, however, are typically rather difficult to extract. Speed-dependent MMs which are not based on wave-making theory, typically do describe continuous dependence of resistance on speed, but do not always accurately predict for the HSC-important position of the main resistance hump (i.e. similar handicap as speed-independent MMs).

Note that for the speed-dependent MMs, speed is usually the most dominant variable, so that care must be exercised to ensure that the other variables are visible. In general, speed-dependent MMs seem to be better for the integrated power prediction approaches.

There is a hybrid approach (Swift et al. 1973; Radojčić 1985). Essentially, with this method the speed-independent equations (with common independent variables for each Froude number) are developed first. A second regression analysis is then performed with the regression coefficients cross-faired against speed (or Froude number). Thus accurate speed-independent equations are obtained for discrete

Froude numbers through the first step, and the second step provides a speed-dependent equation. Of course, either of these equations may be used independently to estimate resistance.

An important and delicate part of this method is the development of the "best subsets" from the initial, for all speeds the same, equation (see discussion on "the best equation" in Sect. 2.2.1). The usual statistical metrics (coefficient of determination, t-test or F-test, standard deviation, significance test for each variable etc.) were found to be insufficient, and in fact were sometimes even misleading. Therefore, a trial and error technique is used to define the best subsets for the whole speed range (see Radojčić 1985). Some variables, judged to be less significant, were rejected deliberately, although a stepwise method was used throughout the analysis. It should be pointed out however, that with this technique, several very good, although dissimilar, models may be derived.

3.3 Systematic Series Applicable to Conventional High-Speed Craft

A systematic or methodical series consists of ship models that are based on a given parent hull, which at the time of testing is usually representative of a state-of-the-art hull form. The principal parameters of these models are obtained by varying their particular dimensions, which results in a systematic change of, for instance, length-beam ratio, deadrise angle, block coefficient etc. There are several systematic series of conventional ships, however only the HSC series are of interest here. Note that hull form, test procedures, or the way results are presented often evolve over time. Consequently, some series may no longer be of interest and may even be outdated.

This section presents a review of the available systematic series applicable to the HSC, upon which the MMs discussed here are based. Monohulls are sorted in chronological order and presented in Table 3.1. Similar information, together with parent body planes, is given in Blount and McGrath (2009).

The catamaran Series 89' is considerably different and is placed at the end of Table 3.1. Series 89' has an additional peculiarity in that it is the only one for which resistance and self-propulsion tests were done. Hence its results are used for extracting MMs for both the resistance and for the power predictions.

Series 62 and DSDS

Series 62 and DSDS (Delft Systematic Deadrise Series) require additional clarification. Namely, Series 62 consists of 5 models with deadrise angle $\beta = 12.5°$. It is tested in DTMB and the results were published in 1963 (Clement and Blount 1963). A sequence of follow-up experiments were carried out in DUT to investigate the influence of deadrise angle. These experiments were performed in several phases over a long period of time. Each phase consisted of models—subseries—where several parameters were systematically changed, except for the deadrise angle (β) which was kept constant.

Table 3.1 Systematic series of HSC for which calm-water resistance and dynamic trim was modeled (for the catamaran Series 89' power too)

Name reference	Main characteristics	Range of main parameters	Remarks
Series 50 Davidson and Surcz (1941) original data. Morabito (2013) data converted to contemporary format.	Hard chine, planing hulls. Warped bottom with negative keel angle and deep forefoot. Consists of 20 models.	$Fn_\nabla = 1.0-6.0$ $L_P/\nabla^{1/3} = 5.3-8.9$ $L_P/B_{PX} = 2.2-8.5$ $LCG/L_P = 0.15-0.43$ $C_B = 0.407$ Interrelated parameters are: $\frac{B_{PX}}{T} = C_B \cdot \frac{(L_P/\nabla^{1/3})^3}{(L_P/B_{PX})^2}$ $\beta_M \approx \tan^{-1}\left(\frac{1.35}{B_{PX}/T}\right)$	Obsolete hull form for planing regimes, nowadays applicable for the semi-displacement regimes. Hull form enables placement of less inclined propeller shafts.
Series 62 (see DSDS) Clement and Blount (1963) $\beta = 12.5°$	Hard chine, narrow-stern, planing hull form. Constant deadrise over the after 50% of hull. Consists of 5 models with $\beta = 12.5°$. $L_P = 1.2-2.4$ m.	$Fn_\nabla = 0.75-6$ $A_P/\nabla^{2/3} = 4.0-8.5$ $L_P/B_{PX} = 2-7$ $LCG/L_P = 0.368-0.488$ $\beta = 12.5°$. $(CA_P - LCG) \cdot 100/L_P$ $= 0-12\%$ $CA_P/L_P = 0.488$; $B_{PT}/B_{PX} = 0.64$ (for wider models a bit different) $L_P/\nabla^{1/3} \approx 1.1$ $(A_P/\nabla^{2/3} \cdot L_P/B_{PX})^{1/2}$	Often analyzed together with DSDS. Extensively investigated series (seakeeping, shallow water, appendages etc.). Used for bench-marking.
Series 65-B (65-A is not of interest here). Holling and Hubble (1974) Hadler et al. (1974)	Hard chine, wide-stern, planing hull form. Small warp over after portion. Three groups of models, in total 9.	$Fn_\nabla = 1.0-4.0$ $\beta_T = 16.3°, 22.5°$ and $30.4°$ $A_P/\nabla^{2/3} = 5.5-8.5$, for lower Fn_∇ from 4 $L_P/B_{PX} = 2.35-9.38$ $LCG/L_P = 0.38$ $CA_P/L_P = 0.397$; $B_{PT}/B_{PX} = 0.99$	Fixed-trim method of testing. Results converted to free-to-trim. Initially developed for hydrofoils (65-A of airplane and 65-B of canard type).
NPL series Bailey (1976)	Important, high-speed, transom-stern, round bilge form for displacement and semi-displacement regimes. Consists of 22 models, $L = 2.54$ m.	$Fn_\nabla = 0.6-3.0$ $Fn_L = 0.3-1.2$ $L/B = 3.33-7.50$ $L/\nabla^{1/3} = 4.5-8.3$ $B/T = 1.75-10.77$ $LCB = 6.4\%L$ aft amidship $C_B = 0.397$ $A_T/A_X = 0.52$ $\beta_T = 12°$	Static wetted surface used for all speeds. Given maneuvering, seakeeping, stability underway, propulsion coefficients, influence of transom wedge and spray rails. Investigated influence of LCB position. Some loll instability at high speeds.

(continued)

Table 3.1 (continued)

Name reference	Main characteristics	Range of main parameters	Remarks
VTT series Lahtiharju et al. (1991)	High-speed, transom-stern, small-bilge radius, semi-displacement form. Modified NPL. Consists of 5 models. One model hard chine (converted from round bilge).	$Fn_\nabla = 0.6–3.8$, hard chine to 5.0 $L/\nabla^{1/3} = 4.86–6.59$ $L/B = 2.75–6.25$ $B/T = 4.39–6.90$ $C_B = 0.397–0.60$ $C_X = 0.57–0.87$ $LCB = 0.436L$ $B_T/B = 0.815$ $A_T/A_X = 0.52$ Tests carried also for 130 and 85% design displacement.	Static wetted surface used for all speeds. Hull form suitable for shallow draught and waterjets. Seakeeping tests.
TUNS series Delgado-Saldivar (1993)	Hard chine, wide-stern, planing hull form. Prismatic hull form (constant deadrise and beam over the after 50% of hull). Consists of 9 models of 1–3.5 kg.	$Fn_\nabla = 0.6–4.0$ $L_P/\nabla^{1/3} = 3.9–6.9$ $A_P/\nabla^{2/3} = 5.2–11.6$ $L_P/B_{PX} = 2.5–3.5$ $\beta_{Bpx} = 12–24$ $LCG/L_P = 0.27–0.38$	Experiments performed in a small university basin only 27 m long. Disadvantage: small models. Match the contemporary hulls
SKLAD series Gamulin (1996)	High-speed, transom-stern, round bilge, semi-displacement form. Built-in stern knuckle and spray rail; forward sections deep V. Consists of 27 models, length 3–6 m.	$Fn_\nabla = 1.0–3.0$ $L/\nabla^{1/3} = 4.5–9.7$ $L/B = 4–8$ $B/T = 3–5$ $C_B = 0.35, 0.45$ and 0.55 $LCB = (8.8–9.2)\%$ L_{DWL} aft amidship $\beta_T = 12°$	Stern-knuckle (chine) 20%L, spray rail 40%L. Good for seakeeping. The afterbody bottom flat and hooked (space for propellers). Given propulsion coefficients, C_A and appendage resistance.
NTUA series Grigoropoulos and Loukakis (1999) resistance Grigoropoulos and Damala (2001) dynamic trim	Double-chine (70%L), wide-stern, semi-displacement form with bottom warp. Consists of 6 models, length above 2 m.	$Fn_L = 0.5–0.9$ $L/B = 4.3–7.5$ $L/\nabla^{1/3} = 6.2–10.0$ $B/T = 3.2–6.2$ $C_B = 0.34–0.54$ $LCB = (12.4–14.6)\%L$ aft amidship $\beta_T = 10°$	Hull form similar to that in Savitsky et al. (1972). May be advantageous for $Fn_L > 0.8$. Focus on resistance, seakeeping, and hull simplicity. Intended for large (above 20 m) high-speed ferries. Wide transom for waterjets.
USCG series Kowalyshyn and Metcalf (2006)	Hard chine, wide-stern, planing hull form. Slight bottom warp and deadrise variation. Consists of 4 models of 135–220 kg.	$Fn_\nabla = 0.6–6.0$ $L_P/\nabla^{1/3} = 5.2–6.1$ $A_P/\nabla^{2/3} = 4.8–9.1$ $L_P/B_{PX} = 3.3–4.6$ $\beta_{Bpx} = 18–21$ $LCG/L_P = 0.37–0.41$	Representative length of wetted surface is L_{KEEL}. Reliable results as models are large. Fn_∇ range up to 6. Contemporary hull form.

(continued)

Table 3.1 (continued)

Name reference	Main characteristics	Range of main parameters	Remarks
DSDS (based on Series 62) Keuning and Gcritsma (1982) $\beta = 25°$ (prismatic hulls). Keuning et al. (1993) $\beta = 30°$ (prismatic hulls). Keuning and Hillege (2017a, b) $\beta = 19°$ (prismatic hulls). $\beta = 25°$ (variable deadrise with warp and rocker).	Hard chine, narrow-stern, planing hull form. $L_P = 1$–1.5 m. (a) Const. deadrise over the after 50% of hull. Consists of 13 models (4 + 5+ 4 with $\beta = 19°$, 25°, and 30°, respectively). (b) Variable deadrise over the after 50% of hull. Consists of 6 models.	(a) $Fn_\nabla = 0.75$–3.0 (for some models to $Fn_\nabla = 4$) $\beta = 19°$, 25° and 30° $A_P/\nabla^{2/3} = 4.0$–8.5 $L_P/B_{PX} = 2$–7 $(CA_P\text{-}LCG) \cdot 100/L_P = 0$–12% $CA_P/L_P = 0.488$; $B_{PT}/B_{PX} = 0.64$ (for wider models a bit different) (b) $\beta = 25°$ $\varepsilon = 10°$ and 20° $\gamma = -1.69°$ to $-4.93°$ $L_P/B_{PX} = 4.1$–7.0 Other parameters are same as for the prismatic hulls.	Parent models of DSDS subseries with $\beta = 19°$, 25°, and 30° are developed from the Series 62 parent 4667-1. Subseries with $\beta = 19°$ was tested in 1996 but released in 2017. Entire database (with the Series 62) consist of 24 models. Subseries with warp and rocker has $\beta_M = 25°$ and $\beta_T = 5°$ or 15° and negative keel angle.
NSS De Luca and Pensa (2017)	Hard chine, wide-stern, planing hull form. Slight bottom warp and deadrise variation. Consists of 5 models of 40–122 kg.	$Fn_L = 0.5$–1.6 $Fn_\nabla = 1.1$–4.2 $L/\nabla^{1/3} = 4.83$–7.49 $L_P/B_C = 3.45$–6.25 $LCG/L_P = 0.33$–0.38 $\beta_T = 14.3°$ $\beta_M = 22.3°$ $A_T/A_X = 0.94$	Contemporary hull form. Suggested to be used with the interceptors. NSS notation: L_{WLD}, B_C and L correspond to L_{WL} dynamic, B_{PX} and L_{WL} static.
Series 89' Müller-Graf (1999) (And several other papers published between 1989 and 2000)	Systematic series of hard chine catamarans. Standard config. consist of 12 models of 3.6–5.1 m. Resistance and self-propulsion tests.	$Fn_L = 0.8$–1.4 $L_{WL}/B_{WLDH} = 7.55$, 9.55, 11.55, 13.55 $\beta_M = 38°$, 27°, and 16° $\delta_W = 0°$, 4°, 8°, 12° $(L_{wedge} = 2.8\%L_{WL})$ $L_P/(\nabla/2)^{1/3} = 6.1$–9.6 $LCB = (0.38$–$0.42)L_{WL}$ $B_T/L_{WL} = 0.167$ $B_{TR}/B_{XDH} = 0.92$ $T_{TR}/T_H = 0.54$—for fully wetted prop. $T_{TR}/T_H = 0.68$—for surface piercing $T_{TR}/T_H = 0.95$—for waterjets	Different hull forms tested (symmetric, semi-symmetric, and asymmetric hard chine and round bilge), afterbodies and propulsion configurations (fully wetted propellers (FP and CPP), SPP and WJ). Tests and results validated for trial conditions.

The parent model of each subseries was based on the Series 62 parent DTMB model 4667-1. The results of the subseries with $\beta = 25°$ and $30°$ were published in Keuning and Geritsma (1982) and Keuning et al. (1993), respectively. The cluster of subseries datasets for $\beta = 12.5°$ and $25°$ (10 models), or $12.5°$, $25°$, and $30°$ (14 models) was named, "Series 62", "PHF", or recently "DSDS", depending on the user/author. Obviously, none is fully correct. This author used the name "Series 62". Then in 1996 the subseries with $\beta = 1\,9°$ was tested (additional 4 models), but these results were released twenty years later in Keuning and Hillege (2017a, b), actually through the DUT website. The same publications included results for 6 variable deadrise models with negative keel angle (i.e. with twist[1] and rocker) also tested in two phases. To summarize, the whole series, released over the period of more than 50 years, consists now of 24 models, all based upon the Series 62, and *now* merit enough to be called DSDS.

3.4   Mathematical Modeling of Resistance and Dynamic Trim for High-Speed Craft

This section addresses the MMs for resistance prediction (and dynamic trim where available) of high-speed craft (HSC's MMs). Similar subjects were addressed earlier, for instance by van Oossanen (1980), Almeter (1993), and others, but these are now outdated and merit an update. Some resistance prediction MMs for high-speed round bilge hull forms were re-evaluated recently, see Sahoo et al. (2011). In the following text 18 MMs are discussed and 10 are recommended as they have not yet been superseded. These MMs are presented in Tables 3.2 and 3.3 which are given later in Sect. 3.7. Some of them form the basis of current computerized resistance prediction software packages.

3.4.1   BK and MBK (Yegorov et al. 1978)

A relatively unusual resistance prediction MM, based on the Soviet hard chine BK and MBK series, was developed in the 1970s. This MM had two main differences from all following MMs. Namely:

- From the outset, both BK and MBK series were based on obsolete hull forms when conceived and model tested in the 1960s and 1970s. BK and MBK series were intended to be used for large semi-planing patrol craft and small planing craft, respectively. Their parent forms were actually similar to the Series 50. BK consisted of even 16 2.1–3.9 m and MBK of 13 1.6–2.4 m long models with

---

[1]Note: *Bottom twist* is equivalent to *bottom warp*. The term *twist* is used by the authors from DUT; for the remainder of this text the term *warp* is used.

Table 3.2 Recommended MMs for evaluation of resistance and dynamic trim for semi-displacement hull forms

Name reference	MMs main characteristics	Target value = f (input values)	Range of input parameters	Remarks
Mercier and Savitsky Mercier and Savitsky (1973) See Sect. 3.4.2	High-speed (Fn$_\nabla$ = 1–2), transom-stern, semi-displacement. MM based on six round bilge series (NPL, Nordstrom, De Groot, SSPA, 64, 63, 62) and a single hard chine series (Series 62), in total 118 hull forms.	$(R_T/\Delta)_{100,000}$ = f $(L_{WL}/\nabla^{1/3}, \nabla/B_X^3, i_e, A_T/A_X)$ for each Fn$_\nabla$ = 1.0, 1.1,...2.0 $R_T/\Delta \neq$ f (LCG) LCG$_{OPT}$ = (2–7)%L$_{PP}$ aft of amidship B$_X$ = B$_{WL}$	Constraints given in graphical format (scatter plots). Approximately: $L_{WL}/\nabla^{1/3}$ = 4–9 i_e = 5–50 A$_T$/A$_X$ = 0–0.8 L$_{WL}$/B$_X$ = 2–7	PT: Regression analysis. I/F: 27/13 terms. SI MM. Corrections for LCG $\neq$ LCG$_{OPT}$ and $R_T/\Delta \neq (R_T/\Delta)_{100,000}$ Discrepancies between measured and calculated, less than 10% for 90% of cases Schoenher's fr. coef. C$_A$ = 0.
VTT Lahtiharju et al. (1991) See Sect. 3.4.5	High-speed, transom-stern, round bilge semi-displacement (NPL, SSPA VTT). (Fn$_\nabla$ = 1.8–3.2) Small draught, large C$_B$. MM based on 65 hull forms. .... High-speed, hard chine (Fn$_\nabla$ = 1.8–3.3). MM based on 13 hull forms.	$(R_T/\Delta)_{100,000}$ = f($\nabla^{1/3}$/T, L/T, B/L, A$_T$/A$_X$, B/T, C$_X$, L/$\nabla^{1/3}$, B^3/∇, Fn$_\nabla$) .... $(R_T/\Delta)_{100,000}$ = f(∇/T^3, $\nabla^{1/3}$/L, L/T, L/$\nabla^{1/3}$, B/L, A$_T$/A$_X$, Fn$_\nabla$) L = L$_{WL}$, B = B$_{WL}$	L/$\nabla^{1/3}$ = 4.47–8.3 B^3/∇ = 0.68–7.76 L/B = 3.33–8.21 B/T = 1.72–10.21 A$_T$/A$_X$ = 0.16–0.82 C$_X$ = 0.567–0.888 LCB = 0.436L = const. .... L/$\nabla^{1/3}$ = 4.49–6.81 L/B = 2.73–5.43 B/T = 3.75–7.54 A$_T$/A$_X$ = 0.43–0.995	PT: Regression analysis. SI MM with 24 and 6 terms for round bilge and hard chine respectively. Static wetted surface used for all speeds; ITTC-57. Given influence of variation of Δ, spray rails and wedges. Predictions of seakeeping. For Fn$_\nabla$ = 1.0–1.8 Mercier and Savitsky; Fn$_\nabla$ > 3.3 Savitsky. Dynamic trim not given.
NPL Radojčić et al. (1997) See Sect. 3.4.7	High-speed, transom-stern, round bilge semi-displacement (Fn$_\nabla$ = 0.8–3.0). SI and SD MMs, for Fn$_\nabla$ = 0.8–3.0 and Fn$_\nabla$ = 1.0–3.0, respectively. MM based on solely on the NPL Series.	$(R_T/\Delta)_{100,000}$, τ, (S) = f(L/B, L/$\nabla^{1/3}$, B/T) L = L$_{WL}$, B = B$_{WL}$	Constraints given as nine simple eq. of inequality type. Approximately: L/B = 3.33–7.50 L/$\nabla^{1/3}$ $\approx$ 4.5–8.3 B/T = 1.76–10.77	PT: Regression analysis. I/F: 27/14 for (R/Δ)$_{100,000}$, 27/15 for τ and 27/13 for (S). S does not depend on speed. SD MMs have 14 × 9 = 126 and 15 × 9 = 135 terms for R/Δ and τ respectively. Propulsion coefficients in Bailey (1982).

(continued)

Table 3.2   (continued)

Name reference	MMs main characteristics	Target value = f (input values)	Range of input parameters	Remarks
SKLAD Radojčić et al. (1999) See Sect. 3.4.8	High-speed, transom-stern, round bilge, semi-displacement (Fn$_\nabla$ = 1.0–3.0). MM based on solely on the SKLAD Series.	C_R, S/$\nabla^{2/3}$, L/$\nabla^{1/3}$, τ = f (L/B, B/T, C$_B$) for each Fn$_\nabla$ = 1.0, 1.25,…3.0 L = L$_{WL}$, B = B$_{WL}$	L/B = 4–8 B/T = 3–5 C$_B$ = 0.35–0.55 L/$\nabla^{1/3}$ = 4.5–9.7 Evaluated from L/$\nabla^{1/3}$ = (L/B)$^{2/3}$(B/T)$^{1/3}$(1/C$_B^{1/3}$)	PT: Regression analysis. I/F: 100/22, 30, 21, and 19 terms for C_R, S/$\nabla^{2/3}$, L/$\nabla^{1/3}$, and τ respectively. All MMs are SI.
NTUA Radojčić et al. (2001) See Sect. 3.4.11	High-speed, double-chine, wide-transom, pre-planing semi-displacement (Fn$_L$ = 0.5–0.9). MM based solely on the NTUA Series.	C_R and τ = f (L/B, L/$\nabla^{1/3}$, B/T, Fn$_L$) (S) = f(B/T, L/$\nabla^{1/3}$) L = L$_{WL}$, B = B$_{WL}$	Odd-shaped 3-D space represented with 8 and 11 inequalities for C_R and τ respectively. Approximately: L/B = 4.3–7.5 L/$\nabla^{1/3}$ = 6.2–10.0 B/T = 3.2–6.2	PT: Regression analysis. I/F: 100/31 and 100/28 for C_R and τ respectively. SD MMs. For (S)—6-term MM. The independent variables and S are related to the static waterline.

Legend: PT—Predictive Technique; I/F—Initial/Final MM; SI—Speed-Independent; SD—Speed-Dependent

length-beam ratio (L$_P$/B$_M$), ranging between 3.75–7.00 and 2.50–3.75, respectively. Tested speed range corresponded to Fn$_\nabla$ = 1.0–4.5. The principal reason for obsolescence is the low transom deadrise and high afterbody warp as β_{TR} = 0° and 5°, and β_M = 12°–18° and 7°–18°, respectively for BK and MBK series. Incomplete calm water resistance results were published in the late 1960s through to the late 1970s, ending with the book by Yegorov et al. (1978).
- A relatively unusual MM for calm water resistance prediction was proposed, using a Taylor series expansion, as a f(C_Δ, LCG, β_M, L$_P$/B$_M$). Therefore, the resistance of a target hull, with a form within the range of either MK or MBK series, had to be expressed via a Taylor's series expansion for each Froude number. Based on the abovementioned, a technique for resistance prediction and form optimization of low deadrise hard chine hull forms is given in Almeter (1988).

3.4.2 Mercier and Savitsky—Transom-Stern, Semi-displacement (Mercier and Savitsky 1973)

The first HSC MM that enabled evaluation of total resistance of transom-stern craft in the non-planing regime (Fn$_\nabla$ = 1–2) built by application of regression analysis was developed by Mercier and Savitsky (1973). Their MM is based on seven transom-stern hull series (see Table 3.2) and focused on the lower speed range of HSC. For higher planing speeds, the Savitsky (1964) method was already in use. Selection of four hull form and loading parameters as input variables [slenderness ratio

Table 3.3 Recommended MMs for evaluation of resistance and dynamic trim for semi-planing and planing hull forms

Reference (series)	MMs main characteristics	Target value = f (input values)	Range of input parameters	Remarks
62 & 65 Radojčić (1985) See Sect. 3.4.4	High-speed, hard chine, planing hull forms. SI and SD MMs for $Fn_\nabla = 1.0$–4.0. MM based on the Series 62 ($\beta = 12.5°$ and 25°). Series 65-B and one model DL-62-A. Total 15 hull forms and 98 loadings.	$(R_T/\Delta)_{100,000}$ and $\tau = f(A_P/\nabla^{2/3}, LCG/L_P, L_P/B_{PA}, \beta)$ $S/\nabla^{2/3}$ and $L/L_P = f(A_P/\nabla^{2/3}, LCG/L_P, L_P/B_{PA}, \beta)$	Constraints given in graphical format (scatter plots). Approximately: $A_P/\nabla^{2/3} =$ 4.25–9.5 $LCG =$ (30–44.8)%L_P $L_P/B_{PA} =$ 2.36–6.73 $\beta =$ 13–37.4; or as 4-D ellipsoid (non-linear eq. replacing 60 bounds)	PT: Regression analysis. I/F: 27/18 terms for both $(R_T/\Delta)_{100,000}$ and τ; I/F: 27/8 terms for both $S/\nabla^{2/3}$ and L/L_P. SI and SD MMs (may be used independently). Corrections for $R_T/\Delta \neq (R_T/\Delta)_{100,000}$. This MM replaces Radojčić (1984).
USCG & TUNS Radojčić et al. (2014b) See Sect. 3.4.13	Planing hull forms with wide transom (match the contemporary hulls). $Fn_\nabla = 0.6$–3.5. The applicability of the USCG Series (consisting of 4 models of 135–220 kg) is extended with TUNS Series (consisting of 9 models of 1–3.5 kg).	$(R_T/\Delta)_{100,000} = f$ $(L/\nabla^{1/3}, Fn_\nabla, L/B, LCG/L, \beta)$ $S/\nabla^{2/3} = f (L/\nabla^{1/3}, Fn_\nabla, L/B, LCG/L)$ $L_K/L = f (Fn_\nabla, LCG/L)$ $L = L_P, B = B_{PX}$ $\beta = \beta_{Bpx}$ $(R_T/\Delta)_{100,000} = R/\Delta$	17 inequalities Approximately: $L/\nabla^{1/3} =$ 3.9–6.9 $L/B = 2.5$–4.7 $LCG/L =$ 0.27–0.41 $\beta = 12$–24	PT: ANN (Regression for L_K/L) Final MM for R/Δ, $S/\nabla^{2/3}$ and L_K/L have 116, 23 and 16 terms respectively. SD MMs. L_K is representative length of wetted surface. Dynamic trim not modeled.
Series 50 Radojčić et al. (2014b) See Sect. 3.4.14	Hulls with a warped bottom, negative keel angle and deep forefoot (enables placement of less inclined propeller shafts). Nowadays applicable for the semi-displacement regimes. MM valid for $Fn_\nabla = 1.0$–5.0. MM based solely on the Series 50.	R_R/Δ, τ_{BL}, $L_M/L_P =$ f $(L_P/\nabla^{1/3}, L_P/B_{PX}, LCG/L_P, Fn_\nabla)$ $S = 0.962 \cdot (L_M/L_P) \cdot L_P \cdot B_{PX}$	34 inequalities Approximately: $L_P/\nabla^{1/3} =$ 5.3–8.9 $L_P/B_{PX} =$ 2.2–8.5 $LCG/L_P =$ 0.15–0.43	PT: ANN and Regression for τ_{BL} & R_R/Δ and L_M/L_P respectively. SD MMs. Final MMs for R_R/Δ, τ_{BL}, and L_M/L_P have 65, 91 and 16 terms respectively. The "hydrodynamic trim" τ is approximately 1° less than the base line trim τ_{BL}

(continued)

Table 3.3 (continued)

Reference (series)	MMs main characteristics	Target value = f (input values)	Range of input parameters	Remarks
Series 62 $\beta = 12.5°–30°$ Radojčić et al. (2017a) Radojčić et al. (2017b) See Sect. 3.4.15	High-speed, hard chine, narrow-stern, planing hulls. ($Fn_\nabla = 1.0–4.0$, to 6 for $\beta = 12.5$) Total 14 hull forms and 280 loading cases with 2880 points out of which 10% are virtual. MM based solely on the Series 62 with $\beta = 12.5°$, $25°$, and $30°$.	$(R_T/\Delta)_{100,000}$ and $\tau = \quad f(A_P/\nabla^{2/3},$ $L_P/B_{PX}, LCG/L_P, \beta,$ $Fn_\nabla)$ $S/\nabla^{2/3}$ and $L_M/L_P =$ $f(L_P/B_{PX}, L_P/\nabla^{1/3},$ $LCG/L_P, Fn_\nabla)$ $A_P/\nabla^{2/3} = f$ $(L_P/\nabla^{1/3}, L_P/B_{PX})$	Multidimensional cuboid. Approximately: $\beta = 12°–30°$ $L_P/B_{PX} = 2–7$ $A_P/\nabla^{2/3} =$ $4.0–8.5$ $LCG/L_P =$ $36.8–48.8$	PT: ANN with multiple and single output. SD MMs. The multiple output MMs for $(R_T/\Delta)_{100,000}$ and τ have 225 terms each, out of these 217 terms are common. Single output MMs for R/Δ and τ are with 190 and 160 terms respectively. For Froude expansion separate regression derived MMs for evaluation of $A_P/\nabla^{2/3}$, L_M/L_P and $S/\nabla^{2/3}$ with 7, 11, and 25 terms respectively.
NSS Radojčić and Kalajdžić (2018) See Sect. 3.4.18	Contemporary planing hull form with wide transom and slight bottom warp. $Fn_L = 0.5–1.6\, Fn_\nabla$ $= 1.1–4.2$ MM is based solely on the NSS.	$(R_T/\Delta)_{100,000}, \tau,$ $S/\nabla^{2/3}, L_{WL}/L_P = f$ $(L_P/\nabla^{1/3}, L_P/B_{PX},$ $LCG/L_P, Fn_\nabla)$ $L_{WL} = L_K$	Constraints given in graphical format (scatter plots) Approximately: $L_P/\nabla^{1/3} =$ $5.0–7.8$ $L_P/B_{PX} =$ $3.45–6.25$ $LCG/L_P =$ $0.33–0.38$	PT: ANN, multiple output. Final MMs for R/Δ, τ, $S/\nabla^{2/3}$ and L_{WL}/L_P have 122 terms each, out of these 113 are common for R/Δ & τ and $S/\nabla^{2/3}$ & L_{WL}/L_P. SD MMs. Representative length of wetted surface is L_{KEEL} Dynamic trim and influence of interceptors not modeled.

Legend: PT—Predictive Technique; I/F—Initial/Final MM; SI—Speed-Independent; SD—Speed-Dependent

($L_{WL}/\nabla^{1/3}$), static beam load coefficient (∇/B_X^3), half-angle of entrance of waterline at bow (i_e), and transom area ratio (A_T/A_X)], was based on the authors' experience, and no statistical correlation analysis was performed. The speed-independent approach was employed as wave-making-resistance-theories for speed-dependent MMs were judged not to be applicable for high-speed transom-stern craft. The speed range covered the main resistance hump, as well as several local humps that occur at lower speeds [in terms $R_W = f(Fn_L)$]—which is generally difficult to model with a single equation.

3.4.3 Transom-Stern, Round Bilge, Semi-displacement (Jin et al. 1980)

Jin et al. (1980) followed the Mercier and Savitsky (1973) approach and replaced Series 62 (hard chine hulls), Series 64 (slender forms), and SSPA (flattened-off afterbody), with the Chinese built round bottom hulls, forming the database of 87 hull forms. The objective was speed-independent modeling of residuary resistance of high-speed round bilge displacement hulls, for Fn_L range 0.4–1.0. After examining the influence of various parameters, Mercier and Savitsky's input variables $L_{WL}/\nabla^{1/3}$, ∇/B_X^3, i_e and A_T/A_X that modeled $(R_T/\Delta)_{100,000}$, were replaced by C_∇, C_P, A_T/A_X, LCB, and i_e and modeled C_R. Note that in both cases the most significant parameter is the ratio of length and displacement, whether expressed as $L_{WL}/\nabla^{1/3}$ or $C_\nabla = \nabla/(0.1L)^3$. Thus, practically only ∇/B_X^3 is replaced with C_P and LCB. It is reported, but not quantified, that accuracy was a bit better than that of Mercier and Savitsky. Two speed ranges were identified which resulted in a suggestion to adjust the hull-form parameters for each of them.

3.4.4 62 & 65—Hard Chine, Semi-planing and Planing (Radojčić 1985)

Calm-water resistance (R/Δ) and dynamic trim (τ) of stepless planing hulls for a wide speed range corresponding to $Fn_\nabla = 1.0$–3.5 (to $Fn_\nabla = 4$ for τ) can be estimated from the speed-independent and speed-dependent mathematical models given in Radojčić (1985). The speed-independent MM is a derivative of the previous work (Radojčić 1984), with an altered number of cases, regression equations etc. This MM (Radojčić, 1985) is based on the systematic Series 62 with deadrise angles (β) 12.5° and 25°, Series 65-B, and a single DL-62-A model. Actually, the data for β = 25°(Keuning and Geritsma 1982) were transformed and added to the existing database of Hubble (1974), hence all models were represented on a unique basis regarding model tow force (horizontal through CG), water density and viscosity, ATTC-1947 friction line, correlation allowance etc.

The MM is a function of four hull-form and loading parameters representative for the planing hulls (see Table 3.3). A "dummy" variable was introduced to separately predict R/Δ and τ using a single equation, while still distinguishing between the two different hull form types, i.e. those resembling Series 62 and Series 65-B hull form respectively. This approach avoided the introduction of a fifth input variable.

Further refinement of this MM involved the introduction of a 4-dimensional ellipsoid (actually a single non-linear equation, see Radojčić 1991) which defined the bounds of applicability of the MM, and hence replaced the use of typical linear bounds, normally expressed through 60 equations. Note that the size of the 4-D ellipsoid is defined by the user, and the use of a smaller ellipsoid results in increased accuracy. The motivation for this novel approach was: (a) the bounds of applicability are often strange and not necessarily logical, and (b) the accuracy of the MM is usually reduced at the outer limits of the database. On the other hand, optimization routines require precise applicability bounds.

3.4.5 VTT—Transom-Stern, Semi-displacement (Lahtiharju et al. 1991)

Lahtiharju et al. (1991) developed a 24-term, speed-dependent resistance prediction equation using a regression analysis based on the round bilge NPL, SSPA, and VTT series. A separate, much simpler, 6-term MM was also developed for hard chine craft. Both equations applied for $(R_T/\Delta)_{100,000}$ for speed range $Fn_\nabla = 1.8$–3.2, while the Mercier and Savitsky MM is recommended for lower speeds and Savitsky's empirical method for higher speeds. The database contained five new systematically developed and experimentally tested models; see Table 3.1. Note that in this work relatively unusual draught based parameters were used as input variables (see Table 3.2). A MM for dynamic trim predictions, however, was not given. Altogether, the Lahtiharju et al. (1991) work is important not only because MMs for resistance evaluation were derived, but also because resistance and seakeeping tests were performed for new high-speed, round bilge systematic VTT series.

3.4.6 PHF—Series 62 (Keuning et al. 1993)

For predictions of resistance (R/Δ), dynamic trim (τ), and rise of center of gravity $(RCG/\nabla^{1/3})$, for the entire Series 62 (termed PHF—Planing Hull Form), i.e. $\beta = 12.5°$, $25°$, and $30°$, are given in Keuning et al. (1993). See discussion on Series 62 and DSDS in Sect. 3.3. This paper also presented experimental results for the Series 62 with $\beta = 30°$, thus augmenting the data with $\beta = 12.5°$ and $25°$ which was given in Clement and Blount (1963) and Keuning and Geritsma (1982), respectively. Input variables were $A_P/\nabla^{2/3}$, LCG, and L_P/B_{PX} and separate equations were given for each, $\beta = 12.5°$, $25°$, and $30°$. These 12-term MMs were both deadrise-independent and speed-independent, which required interpolation for intermediate β values between the measured points of $\beta = 12.5°$, $25°$, and $30°$, not to mention 10 discrete Froude numbers ranging between $Fn_\nabla = 0.75$ and 3.0. Moreover, R/Δ for displacements of 5 and 50 m^3 was given, which required further interpolations. MMs for evaluation of length of wetted area and wetted area however were not given. Additional 7-term polynomial equations for evaluation of the effect of warped bottom, i.e. evaluation of the differences in resistance $(\Delta\ R/\Delta)$, dynamic trim $(\Delta\ \tau)$, and sinkage $(\Delta\ RCG/\nabla^{1/3})$, compared to parent model of $\beta = 25°$, were also developed using two new input variables [centerline inclination angle (γ) and warp angle[2] (ε)]. The PHF MM is recently superseded by DSDS MM; see Sect. 3.4.16.

3.4.7   NPL (Radojčić et al. 1997)

Three groups of MMs based on high-speed, round bilge, NPL series and a broad speed range of $Fn_\nabla = 0.8$–3.0 were derived in Radojčić et al. (1997). These included speed-independent and speed-dependent models for resistance $((R/\Delta)_{100,000})$ and dynamic trim (τ), and one for wetted surface coefficient $((S) = S/\nabla^{2/3})$; see Table 3.2. Note that these MMs are based on the NPL hull-form series exclusively, as opposed to the previous works (Mercier and Savitsky 1973; Lahtiharju et al. 1991) which combined the NPL series with the data of other series. Furthermore, since $(R/\Delta)_{100,000}$, τ, and (S), were chosen as dependent variables, data for model-size wetted area and $R_R/\Delta = f(L/\nabla^{1/3}, Fn_\nabla, L/B)$ (see Bailey 1976) were transferred to a new format $(R/\Delta)_{100,000}$, τ, $(S) = f(L/B, L/\nabla^{1/3}, B/T)$. Through the first step, 12 27-term speed-independent equations were formed for $(R/\Delta)_{100,000}$ and τ (one for each Fn_∇ ranging from 0.8 to 3.0) and one for (S) as S does not depend on speed. In the next step, the speed-dependent models for $(R/\Delta)_{100,000}$ and τ were developed through cross-fairing of resistance and dynamic trim regression coefficients against Fn_∇ (as previously done in Radojčić, 1985). Satisfactory results were obtained when Fn_∇ was raised to the eight power, resulting in 126- and 134-terms MMs for $(R/\Delta)_{100,000}$ and τ respectively. Speed-independent models are valid for Fn_∇ range between 0.8 and 3.0, while speed-dependent ones were valid for $Fn_\nabla = 1.0$–3.0, due to instability between $Fn_\nabla = 0.8$–1.0.

3.4.8   SKLAD (Radojčić et al. 1999)

A mathematical representation of both calm water resistance and dynamic trim for the systematic round bilge, transom-stern, semi-displacement SKLAD series (Gamulin 1996) for a speed range of $Fn_\nabla = 1.0$–3.0 was presented in Radojčić et al. (1999); see Table 3.2. The dependent variables are the residuary resistant coefficient (C_R), wetted surface coefficient ($S/\nabla^{2/3}$), length-displacement ratio ($L/\nabla^{1/3}$), and dynamic trim (τ), while the independent variables are L/B, B/T, and C_B. All initial polynomial equations started out with 100 terms, while the final MMs for C_R, $S/\nabla^{2/3}$, $L/\nabla^{1/3}$, and τ had 21, 29, 20, and 18 terms respectively; one for each of 8 discrete Fn_∇ ranging from 1 to 3. Consequently, all derived equations were speed-independent. At the time, considerable effort was made to develop speed-dependent MMs, but without success.

---

[2]Usual measure for longitudinal variation of deadrise is *warp rate*, which is change of deadrise over longitudinal length equivalent to chine beam, and is expressed in *degrees per beam* (see Savitsky 2012; Blount 2014). DUT use *warp angle*, see list of Symbols.

3.4.9   Round Bilge and Hard Chine (Robinson 1999)

The results of regression analysis (WUMTIA regression per Molland et al. 2011) of 30 and 66 models of round bilge and chine hull forms, respectively, tested at the Wolfson Unit over some 30 years, is presented in Robinson (1999). The hull forms are not of systematic form, but do broadly fit into the round bilge or hard chine hull-form category, and thus this database is classified as random hulls. Relying on the fact that a relationship between C-Factor (actually C^2) and relative speed and specific total resistance ($v/L^{1/2} \cdot \Delta/R_{T)}$ does exist, a relatively unusual and obsolete method, set up on the so called C-Factor, was applied ($C = 30.1266 \cdot v/(L)^{1/4} \cdot (\Delta/2P_E)^{1/2}$). These results were valid for $Fn_\nabla = 0.5$–2.75 and the initially trim-optimized, but not specified, conditions (i.e. LCG position was shifted or wedge/trim-tabs were used). The dependent variable was the C-Factor, and the independent variables for chine and round bilge 7-term and 10-term speed-independent equations were $L/\nabla^{1/3}$ and L/B, and $L/\nabla^{1/3}$, L/B, and S/L^2, respectively. Simple 3-term MMs for wetted area were also derived for chine craft (for 6 discrete values of Fn_∇) and displacement craft (for static condition only), and depend on Δ, L, and B. According to Molland et al. (2011) the abovementioned MMs overestimate power by some 3–4%.

3.4.10   Transom-Stern, Round Bilge (Grubišić and Begović 2000)

Resistance data of 12 fast round bilge systematic series with 186 models were analyzed in Grubišić and Begović (2000). In addition to the aforementioned round bilge series, a fast twin-screw displacement ship series and a semi-planing series given in Compton (1986) were added. Using regression analysis, a 12-term speed-independent MM was defined with 7 independent variables. These are: slenderness ratio $L/\nabla^{1/3}$, beam draught ratio B/T, prismatic coefficient C_P, maximum area coefficient C_X, transom area coefficient A_T/A_X, Taylor wetted surface coefficient C_S, and longitudinal center of buoyancy L_{CB}/L_{WL}. The dependent variable is residuary resistance coefficient C_R. This MM is valid for the speed range that corresponds to $Fn_L = 0.3$ to 1.2 and is essentially similar to Mercier and Savitsky (1973) and Fung (1991). It is a random hull form type MM as is based on similar round bilge hulls and is usable in the concept design phase.

3.4.11   NTUA (Radojčić et al. 2001)

In Radojčić et al. (2001) raw model test data of the NTUA Series was used for the development of the MMs for resistance (C_R) and dynamic trim (τ), hence regression analysis was applied for both, fairing of the raw data and actual model extraction.

Modeling of the NPL and SKLAD series did not require this since the data were already faired. The speed-dependent approach for modeling of C_R and τ was chosen with L/B, $L/\nabla^{1/3}$, B/T, and Fn_L being the independent variables. Note that two of the independent variables defined the third one, so there was no need to use all three; however all three were incorporated (even though that was wrong from the statistical point of view), because (a) this simplified the mathematical models and, (b) it prevented the use of the MM for cases that are very dissimilar to the NTUA series. This is de facto a new modeling approach applied for resistance and trim predictions.

3.4.12   Displacement, Semi-displacement, and Planing Hull Forms (Bertram and Mesbahi 2004)

ANN as an extraction tool was used for derivation of simple equations for resistance and dynamic trim evaluation (see Bertram and Mesbahi 2004). The MMs derived are based upon an extremely wide hull form spectrum, ranging from the fast displacement ships to hard chine planing hulls. This means that the secondary hull form parameters are neglected. Moreover, oversimplified formulas and diagrams underlie the dataset, which inherently resulted in neglecting some important parameters (e.g. L/B, which is usually one of the primary parameters). A nonlinear relationship between the input variables and an output is derived by a simple ANN with a single hidden layer. For evaluation of total resistance coefficient $C_{TV} = R_T/(\rho/2 \cdot v^2 \cdot \nabla^{2/3}) = f(Fn_L, C_V)$ three groups of equations were derived, covering speed (Fn_L), and slenderness ratios ($C_V = \nabla/L^3$), of 0.2–1.2 and 0016–0.007, respectively. Three equations for evaluation of appendage resistance $R_{APP}/R_T \% = f(Fn_L)$ were also derived, for configurations with 2, 3, and 4 propellers, and two for dynamic trim $\tau = f(\nabla^{2/3}/B \cdot T)$, for semi-displacement and planing hull forms. Resistance, appendage resistance, and dynamic trim equations have on average 20, 10, and 15 terms, respectively. To summarize, an overly sophisticated extraction tool was used for an oversimplified approach (where $R_T = f(Fn_L, C_V)$).

It should be noted however, that this is probably the first application of ANN to HSC monohulls. ANN modeling of catamaran resistance was described in Couser et al. (2004), and Mason et al. (2005), also mentioned in Sect. 2.2.2, but this topic is not discussed further because an adequate MM for resistance prediction has not been released (see Sect. 1.5.3).

3.4.13   USCG & TUNS—Hard Chine, Wide Transom, Planing (Radojčić et al. 2014a)

A mathematical representation of resistance for wide-transom planing hull forms based on the USCG and TUNS Series (see Kowalyshyn and Metcalf 2006; Delgado-Saldivar 1993, respectively) for the speed range corresponding to $Fn_\nabla = 0.6–3.5$ is presented in Radojčić et al. (2014a). Regression analysis and Artificial Neural Network (ANN) techniques are used to establish, respectively, "Simple" and "Complex" mathematical models. For the Simple model, the dependent variable was $(R/\Delta)_{100,000}$, while Fn_∇ and $L/\nabla^{1/3}$ were chosen as the independent variables (being the two dominant high-speed parameters). For the Complex model, additional independent variables were L/B, LCG/L, and β. Both types of MMs are obviously speed-dependent. Being a 2-parameter formulation, the Simple model is intended for use during the concept design phases where the reduced quality of resistance predictions is acceptable. The Complex MM is intended for use for various performance predictions during all design phases. Relatively simple MMs for wetted surface $(S/\nabla^{2/3})$ and its length (L_K/L) were also developed, without β and even without $L/\nabla^{1/3}$, L/B, and β as dependent variables, respectively. Dynamic trim however, could not be modeled. The principal disadvantage of all the derived models is that the TUNS Series consists of very small models, which to some extent introduces unreliability (see Moore and Hawkins 1969; Morabito and Snodgrass 2012; Tanaka et al. 1991 regarding the usefulness of small models).

The Simple model for $(R/\Delta)_{100,000}$ was developed in two phases: (a) Data having same $L/\nabla^{1/3}$ were grouped (regardless of the other hull form and loading parameters), then a trend line $R/\Delta = f(Fn_\nabla)$ was produced for each group, and (b) A second regression analysis is then performed with the regression coefficients cross-faired against the slenderness ratio. The Complex speed-dependent version, derived with an ANN routine, for $(R/\Delta)_{100,000}$ and $S/\nabla^{2/3}$ have 116 and 23 equation terms, respectively. The MM for L_K/L has 16 terms only and was derived with regression analysis.

Comparing the methods used (tools for MM development) ANN showed to be a very good modeling tool for resistance predictions. Regression analysis required more time and higher levels of skill, at least for complex relations with many polynomial terms; it seems to be more convenient for simpler relations/equations.

3.4.14   Series 50 (Radojčić et al. 2014b)

Further assessments of contemporary ANN and conventional regression analysis for modeling of resistance were investigated in Radojčić et al. (2014b). Mathematical representations for predicting resistance (this time R_R/Δ), dynamic trim (baseline trim τ_{BL}), and wetted length (L_M/L_P), of the EMB Series 50 are given in the same reference. The Series 50 database consisted of a re-analyzed full data set, as discussed in

Morabito (2013), because the original model-test data (Davidson and Suarez 1941) were prepared before contemporary planing hull coefficients were introduced. Models derived by regression analysis were somewhat "stiffer" than ANN models. This is not only due to the smaller number of terms, but also to the different format, as regression used polynomials while ANN used a much more complex nonlinear function. The "double hump" phenomenon for dynamic trim, between Fn_∇ 2.0 and 3.0, often connected with dynamic instability, was also noted.

$L_P/\nabla^{1/3}$, L_P/B_{PX}, LCG/L_P, and Fn_∇ were used as the independent variables throughout the work and hence the MMs were speed-dependent. The dependent variables were R_R/Δ, τ_{BL}, and L_M/L_P. Three mathematical models, finalists amongst the several hundred tested, were developed for each of the dependent variables. The final MMs for resistance and trim evaluation were 65-term and 91-term equations derived by regression and ANN, respectively. MMs for predicting L_M/L_P were much simpler, so there was no need to use ANN as the regression technique sufficed.

A detailed quantification of the differences between ANN and regression methods for developing stable and accurate MMs is provided. ANN has been demonstrated as a viable technique for fitting complex data sets accurately—as required for modeling HSC resistance. Modeling dynamic trim is even more challenging and ANN-derived-models successfully replicated the double hump—something that was not achieved with the regression analysis approach.

3.4.15 Series 62 (Radojčić et al. 2017a, b)

After the transition period—from regression to ANN (previous two references)—the author and his team were encouraged to continue application of ANN for similar problems. Follow resistance and trim modeling of the well-known planing hull Series 62 (Radojčić et al. 2017a, b). When this work was performed Series 62 (or DSDS, see discussion in Sect. 3.3) consisted of three groups of experiments, each with a deadrise angle of $\beta = 12.5°$, $25°$, and $30°$, conducted across three decades; see Table 3.1 and Clement and Blount 1963, Keuning and Geritsma 1982, and Keuning et al. 1993, respectively. The first two groups of models ($\beta = 12.5$ and 25) were also used in Radojčić (1985), while all three in Keuning et al. (1993), where derived equations were both deadrise-independent and speed-independent.

Consequently, the original experimental data given in the abovementioned references was rearranged and used for formation of a revised database (Radojčić et al. 2017a) which consisted of original and 'virtual' measurements. The virtual measurements were introduced to cover zones which were not sufficiently covered by the experimental data, ensuring the polynomial fits were fair and continuous between data points. Then, stable speed- and deadrise-dependent MMs for R/Δ and τ predictions for Fn_∇ and β range $1°-4°$ and $12.5°-30°$ respectively, were extracted. The independent variables were β, L_P/B_{PX}, $A_P/\nabla^{2/3}$, LCG/L_P, and Fn_∇, and the dependent ones were R/Δ and τ.

It should be noted that a novel *multiple output* ANN technique was applied, in contrast to *the single output* technique (where resistance data was used only for modeling of R/Δ, and separately, trim data was only used for modeling of trim (τ), hence the structure of ANN models are inherently dissimilar, yielding entirely different equations for R/Δ and τ). With the *multiple output* ANN, all available R/Δ and τ data are used simultaneously, producing slightly different equations for R/Δ and τ. This implied that τ data influenced the model for R/Δ and vice versa, which makes sense in the high-speed regime where R/Δ and τ curves mirror each other. The multiple-output ANN technique resulted in an equation with 233 terms defining both R/Δ and τ values, with only 8 terms differing between them (i.e. 217 terms are common, 225 terms define each, R/Δ and τ). Independently derived, single-output MMs for R/Δ and τ include 190 and 160 terms respectively. The multiple output approach is a novel ANN application for this kind of problems. Single output models, however, have been found to have a bit better accuracy, which is to be expected since they track independently either R/Δ or τ.

3.4.16   DSDS (Keuning and Hillege 2017a, b)

MMs for DSDS (Delft Systematic Deadrise Series) consist of two different sets of MMs developed by the same authors a month apart—Keuning and Hillege (2017a, b). Each MM is for resistance and dynamic trim prediction. Both are derived by regression analysis and are similar to MMs for PHF given in Keuning et al. (1993), with the exception that the new MMs are based on the results of the entire DSDS subseries, and hence include $\beta = 19°$ as well. Specifically, MMs (separate equations for each deadrise angle) for evaluation of R_T/Δ, $\tau = f(A_P/\nabla^{2/3}, L_P/B_{PX}, LCG)$, are given in the first paper. They are both deadrise- and speed-independent—as in Keuning et al. (1993), 24 years earlier. A more advanced approach is given in the second paper with a set of reformatted equations, i.e. R_R/Δ, τ, $L_M/\nabla^{1/3}$, $S/\nabla^{2/3} = f(\beta, A_P/\nabla^{2/3}, L_P/B_{PX}, LCG)$. Since β is among the input parameters, these equations are now deadrise-dependent, but are still speed-independent. These polynomial equations have 17 and 8 terms for evaluation of R_R/Δ & τ and $L_M/\nabla^{1/3}$ & $S/\nabla^{2/3}$ respectively.

Two sets of equations for warped hulls follow the logic and format of prismatic hull equations, and hence depend on the same input variables as the prismatic hulls, in addition to warp and buttock angles (ε and γ, respectively). Note that the PHF equations for warped hulls (Keuning et al. 1993) did not depend on L_P/B_{PX}. The new MM now does—owing to the extension of the warped hull subseries (presently it consists of six models having L_P/B_{PX} range between 4.1 and 7.0). The warped hull equations are essentially evaluating corrections dR_R/Δ, $d\tau$, $dL_M/\nabla^{1/3}$, and $dS/\nabla^{2/3}$ relative to a prismatic hull with $\beta = 25°$. It is assumed that the same correction may be applied to other deadrises.

3.4.17 NSS (De Luca and Pensa 2017)

Model test results in calm water for the NSS (Naples Systematic Series) have been presented in a recent paper by De Luca and Pensa (2017) for dynamic trim (τ), total and residuary resistance coefficients (C_T and C_R), wetted surface (S), and waterline length (L_{WL}); see Table 3.1. NSS is envisaged to be used with the interceptors, but interceptor effect is yet to be published. In the referenced work, for each of the five NSS models and LCG/L_P = 0.38, regression analysis was used to derive a 20-term polynomial equation for evaluation of resistance (actually C_R), running wetted surface, and waterline length as a function of $L_P/\nabla^{1/3}$ and Fn_L. These MMs are speed-dependent, but L_P/B_{PX} and LCG independent. The equations actually evaluate C_R, S, and L_{WL} within the tested $L_P/\nabla^{1/3}$ range, but do not interpolate between the two LCG positions and the various L_P/B_{PX} values. This is similar to the "Simple" 2-parameter MM of Radojčić et al. (2014a) except that each of the five polynomial equations are valid for a particular NSS model, rather than for the whole series.

3.4.18 NSS (Radojčić and Kalajdžić 2018)

Radojčić and Kalajdžić (2018)[3] fill in the gaps in the abovementioned simplified 2-parameter NSS MM and present MMs of resistance (actually $(R_T/\Delta)_{100,000}$), dynamic trim (τ), wetted area ($S/\nabla^{2/3}$), and length of wetted area (L_{WL}/L_P), as functions of L_P/B_{PX}, $L_P/\nabla^{1/3}$, LCG/L_P, and Fn_∇. An Artificial Neural Network (ANN) method with multiple outputs is used to develop the enhanced mathematical models enabling simultaneous use of all the available R/Δ and τ data on one side, and $S/\nabla^{2/3}$ and L_{WL}/L_P on another. Moreover, the ANN structures for both datasets, R/Δ & τ, and $S/\nabla^{2/3}$ & L_{WL}/L_P, are identical, thereby further simplifying the programming. Two equations with 122 terms define each, R/Δ & τ and $S/\nabla^{2/3}$ & L_{WL}/L_P, with even 113 terms in common for R/Δ & τ and for $S/\nabla^{2/3}$ & L_{WL}/L_P. This demonstrates the relationship between dynamic trim and resistance, and wetted surface and its length, for planing craft. Once the influence of the interceptors is revealed, new MMs can be developed with, for example depth of interceptor, as a possible fifth independent variable.

3.5 Future Work—Stepped Hulls

Speeds corresponding to Fn_∇ of up to 8 are relevant for the small planing craft. Stepped hull forms would be advantageous for these speeds (not only from the resistance, but also from the dynamic stability viewpoint). For these hull types some

---

[3]This is an upgraded and corrected version of work published under the same title at the High Speed Marine Vehicles Conference (HSMV 2017) in Napleas 2017.

experimental data exist (for instance, Taunton et al. 2010; De Marco et al. 2017), although adequate MM for resistance and dynamic trim evaluations are missing (except some tries based on the Savitsky method).

3.6 Mathematical Model Use

MM produce incorrect results for two main reasons:

1. MM is not good enough, as

    (a) It does not satisfactorily represent the experimental results it is based on, and/or
    (b) There is an unexpected behavior between the original data points.

2. MM is used incorrectly (discussed also in Sect. 1.4), as

    (a) Boundaries of applicability are violated, and/or
    (b) MM is not applicable for the target hull form.
         (Note that this is same as when inadequate prototype or wrong experimental results are used for the target hull's resistance evaluation).

The MM-developer is responsible for the errors of the first type, while MM-user is responsible for those of the second type. This should be taken into consideration when the MM's quality is judged. In fact, the only truly correct metric of MM's quality is a comparison of predicted vs. measured value (errors ad 1.a above). Evaluation of the predicted model values that lie between the measured data points (ad 1.b) is not trivial and usually is performed by the MM developer.

The cause of the errors in 2.a is obvious, since the boundaries of MM applicability must be obeyed. However, the errors in 2.b require further discussion. Namely, just because the input parameters of a target hull are within the applicability boundaries of a MM, it does not mean that the MM is really usable for the particular target hull (discussed also in Sect. 2.1.1). That is, the target hull's secondary characteristics (i.e. hull form) may be dissimilar from those upon which the MM is based. For instance, the input parameters of a target semi-displacement yacht (L/B, $L/\nabla^{1/3}$ etc.) may satisfy the applicability boundaries of a MM for trawlers, but the hull form of these two vessel types are considerably different, which obviously disqualifies this MM.

Nevertheless, there are cases when the available MM is developed for the same or similar vessel types as the target hull, but the secondary parameters of the respective hulls differ (e.g. MM is based on the narrow stern planing hull forms, while the target hull is of a wide transom form). In these cases the MM may be applied, but the target hull's input parameters required by the MM should be modified. This modification is called "mapping of input parameters" and consists of replacing the input parameters with the suitably adjusted *effective* values. For instance, the required input parameter could be L_P/B_{PA} and/or β_M, while the effective ones could be L_P/B_{PX} and/or β_{Bpx}, respectively. Consequently, the effective input parameters are not necessarily the same as the ones suggested for the MM. Mapping is not a straightforward procedure because an understanding of the issue is necessary and hence, in most cases is either ignored or is wrongly employed. No wonder then that mapping is amongst the principal causes of errors (see Blount and Fox 1976; Savitsky and Brown 1976; Savitsky 2012, etc.). Note that mapping is much more important (influential) for dynamic trim prediction than for resistance prediction (see Radojčić et al. 2017b).

It follows that the MM's users should be aware of the hull forms and development methods used for derivation of the MM. Moreover, without adequate interpretation of the input parameters and prediction results, statistically based MMs can lead to erroneous results. MacPherson (2003) summarized typical problems with numerical model performance predictions and proposed practical solutions. Similarly, regardless of how resistance and propulsion are estimated (i.e. by model experiments, analytical method, CFD etc.), common powering prediction errors are given in Almeter (2008). Over fifty possible sources of errors are discussed with a focus on the design. It is concluded that the most accurate prediction method is not always the most expensive one.

A MM's predictive accuracy is of paramount importance for the user; see Sect. 2.1. However, it is not really possible to determine the accuracy of the MM's predictions even when the developer took all possible precautions and checked the relevant statistics, assessed the discrepancies between measured and evaluated values, and validated the behavior of the MM between the data points. A "correlation procedure" which could be used to improve a MM's accuracy is given in van Hees (2017). This approach essentially consists of tuning the MM to hull forms similar to the subject hull, but with known and available performance results. The same paper clarifies the correlation[4] procedures applying the Holtrop and Mennen method (Holtrop and Mennen 1982). Note that a similar approach, i.e. correlation procedure, could be used for improving the mapping.

The predictions of the four MMs *based* on the Series 62 are compared with the Series 62 (or DSDS) measurements in Figs. 3.1 and 3.2. A comparison of predicted results with the test results of *similar* hull forms is more demanding; see Figs. 3.3 and 3.4. Note that a user can be sure of the results only by using as many predictive methods as possible. That is, in everyday design application the user has to rely entirely on the MM's predictions as the actual value is not known.

3.7 Recommended Mathematical Models for Resistance and Dynamic Trim Prediction

The following MMs for resistance and dynamic trim prediction, out of the 18 that have been discussed here, are recommended:

---

[4] "Correlation is the process of comparing experimental and numerical results in ship hydrodynamics with the aim to improve prediction accuracy" (van Hees 2017).

Fig. 3.1 Comparison of four MMs with the Series 62/DSDS measurements—β = 25°, $A_P/\nabla^{2/3}$ = 7, %LCG = 4%, L_P/B_{PX} = 4.1 (single- and multiple-output—Radojčić et al. 2017b, DSDS—Keuning and Hillege 2017a, 62 & 65—Radojčić 1985)

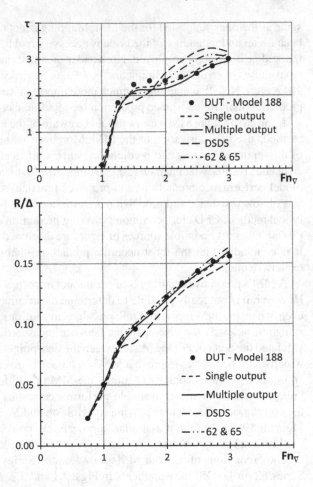

MMs for Semi-displacement Hull Forms

Transom-stern Mercier and Savitsky (MM of random hull form class) and NPL, VTT, SKLAD, and NTUA (methodical series class MMs). The length Froude number (Fn_L) of up to 1.1 is of interest for all of them. Applicability range of these MMs is shown in Fig. 3.5.

MMs for Semi-planing or Planing Hull Forms

62 & 65 and USCG & TUNS (each MM comprises two methodical series, and are therefore a kind of random hull form class), and 50, 62, and NSS (methodical series class MMs). For this group volumetric Froude number (Fn_∇) of up to 5.5 or so is appropriate. Applicability range of these MMs is shown in Fig. 3.6.

For quick reference the main characteristics of the recommended MM are presented in the Tables 3.2 and 3.3.

Fig. 3.2 Comparison of four MMs with the Series 62/DSDS measurements—β = 12.5°, $A_P/\nabla^{2/3}$ = 5.5, %LCG = 4%, L_P/B_{PX} = 4.1 [single- and multiple-output—Radojčić et al. (2017b), DSDS—Keuning and Hillege (2017a), 62 & 65—Radojčić (1985)] (Note: In the τ = f(Fn$_\nabla$) graph, three data points for Fn$_\nabla$ = 1.75, 2.00 and 2.25 are outliers, i.e. they are measurement errors, according to Radojčić et al. (2017b). This was detected by ANN based MMs, while regression based MMs obviously tracked erroneous measurements)

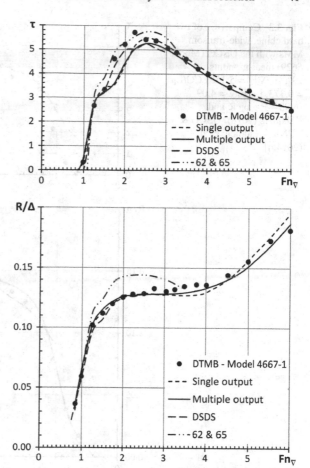

Note for Figs. 3.5 and 3.6: Dimensionless speed and loading parameters are interrelated, i.e. $Fn_\nabla = Fn_L \cdot (L/\nabla^{1/3})^{1/2}$ and $L_P/\nabla^{1/3} \approx 1.1 \cdot [A_P/\nabla^{2/3} \cdot L_P/B_{PX}]^{1/2}$ (assuming $L/L_P \approx 0.98 \approx 1$ and $A_P \approx 0.83 \cdot L_P \cdot B_{PX}$, which stems from the Series 62 with average value for $B_{PX}/B_{PA} \approx 1.22$); see Blount (2014).

MMs for Series 62/DSDS

Additional elaboration is required for the MMs for Series 62/DSDS (see also discussion on Series 62 and DSDS in Sect. 3.3) as there are four of them. The author's opinion follows:

- MM 62 & 65 (Radojčić 1985) is the oldest. Nevertheless, in some cases it still might be useful since it contains some of Series 65-B characteristics. It should be used well within its own boundaries of applicability.

Fig. 3.3 Comparison of two hard chine, wide-transom MMs with the USCG (Model 5629) measurements—β = 23°, $L_P/\nabla^{1/3}$ = 5.15, LCG/L_P = 0.373, L_P/B_{PX} = 4.09 [NSS—Radojčić and Kalajdžić (2018), USCG & TUNS—Radojčić et al. (2014a)]

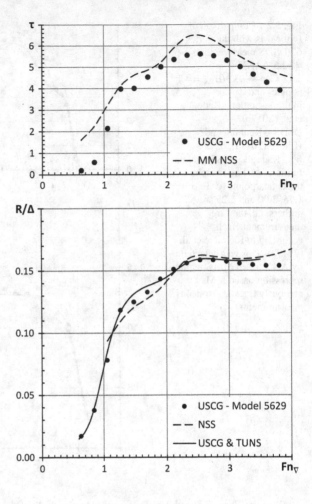

- PHF MM (Keuning et al. 1993) was not practical even when released (speed- and deadrise-independent, for 5 and 50 m³ only). It is superseded by MMs released in Keuning and Hillege (2017a, b).

- Two new versions of DSDS MMs for prismatic (constant deadrise) hulls have been released recently (Keuning and Hillege 2017a, b; regression coefficients are available on DUT's website). The second version is more advanced than the first, even though it is still speed-independent. Both MMs are based on the subseries with β = 19°, which should be an advantage. However, since these experimental results seem to be incorrect, these MMs may be invalid too, at least for β = 19°. This conclusion was reached when (a) β=19° results were compared with predictions of other MMs, and (b) β = 19° data was correlated with the data of other "siblings" (DSDS subseries with β = 12.5°, 25°, and 30°). MMs for warped hulls, however, may be useful.

Fig. 3.4 Comparison of two hard chine, wide-transom MMs with the NSS (Model C2-T13) measurements β = 22.3°, $L_P/\nabla^{1/3}$ = 5.95, LCG/L_P = 0.38, L_P/B_{PX} = 3.89 [NSS—Radojčić and Kalajdžić (2018), USCG & TUNS—Radojčić et al. (2014a)]

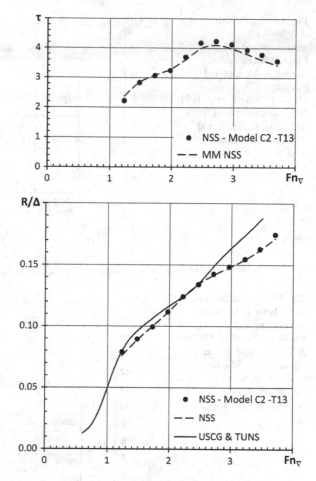

- Series 62 MM (Radojčić et al. 2017b), is by far the most advanced as is a f($A_P/\nabla^{2/3}$, L_P/B_{PX}, LCG/L_P, β, Fn_V). It seems to be the most accurate and reliable, especially in the intermediate range (between the data points). Lack of β = 19° data in its foundation is substituted with the so called "virtual measurements" (see Radojčić et al. 2017a). Therefore, for the time being (mid 2018), this MM is recommended.

It is expected, however, that once a reliable database is available, a new and probably final Series 62/DSDS MM will be developed, and will probably include the characteristics for the DSDS subseries for warped hull forms. In any case, contemporary MMs should be both deadrise- and speed-dependent.

Fig. 3.5 Applicability range
of MMs for
semi-displacement hull
forms (Mercier and
Savitsky—Mercier and
Savitsky 1973,
VTT—Lahtiharju et al. 1991,
NPL—Radojčić et al. 1997,
SKLAD—Radojčić et al.
1999, NTUA—Radojčić
et al. 2001)

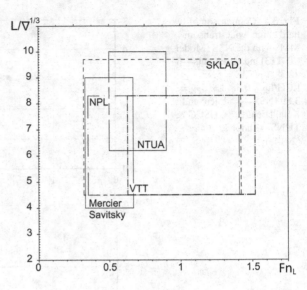

Fig. 3.6 Applicability range
of MMs for semi-planing
and planing hull forms (62 &
65—Radojčić 1985, USCG
& TUNS—Radojčić et al.
2014a, Series 50—Radojčić
et al. 2014b, Series
62—Radojčić et al. 2017b,
NSS—Radojčić & Kalajdžić
2018)

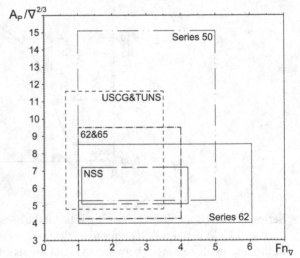

3.7.1   Some Typical Examples

Resistance and dynamic trim predictions of five MMs applicable for semi-displacement hull forms (4 round bilge, 1 double-chine) are shown in Fig. 3.7. The hull dimensions used are typical for a 500 t mega yacht (L_{wl} = 52 m, B_{wl} = 9.3 m, T = 2.55 m, LCG = 23.4 m), and hence the main non-dimensional parameters are $L_{wl}/\nabla^{1/3}$ = 6.6 and L_{wl}/B_{wl} = 5.6 (as per Blount 2014). Other assumed variables used as input for some MMs, were: A_T/A_X = 0.52 and C_B = 0.396 according to the NPL

Fig. 3.7 The Predictions of five MMs (4 round bilge, 1 double-chine—NTUA) applicable for a typical 500 t mega yacht (L_{wl} = 52 m, B_{wl} = 9.3 m, T = 2.55 m, LCG = 23.4 m)

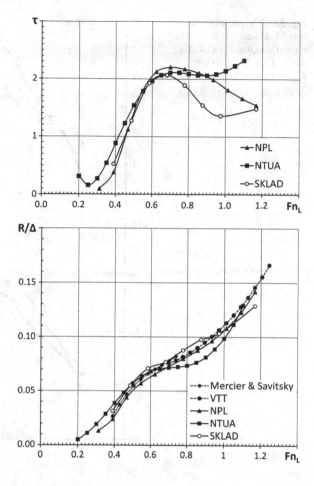

series, $C_X = 0.573$ according to the VTT series, and $i_e = 15°$ an average value. R/Δ is valid for $C_A = 0$ and 500 t, hence here R/Δ = (R/Δ)$_{500t}$.

Predictions of four MMs applicable for hard chine hull forms are shown in Fig. 3.8. The hull dimensions are typical for a 100,000 lb (45.4 t) yacht (L_P = 23 m, B_{PX} = 4.1 m, LCG = 8.6 m), and hence the main non-dimensional parameters are $L_P/\nabla^{1/3} = 6.5$ and $L_P/B_{PX} = 5.6$ (as per Blount 2014). $C_A = 0$ for all cases.

Additional elaboration is needed for the Series 62 MM because the input parameters are not the same as for other MMs. That is, (a) the planing area coefficient $A_P/\nabla^{2/3}$ is used instead of the slenderness ratio $L_P/\nabla^{1/3}$ (relationship $A_P/\nabla^{2/3} = f(L_P/\nabla^{1/3}, L_P/B_{PX})$ is given above), and (b) in order to cover the deadrise span of other series, two Series 62 curves are evaluated, for β = 16° and 22.5°; see Fig. 3.8.

Note that the MM for Series 50 predicts the baseline trim (τ_{BL}). Hydrodynamic trim (shown in Fig. 3.8) is obtained as $\tau_{BL} - 2°$ (for this particular example 2° static trim is assumed).

Fig. 3.8 The predictions of
four MMs applicable for a
typical 100,000 lb (45.4 t)
yacht (L_P = 23 m, B_{PX} =
4.1 m, LCG = 8.6 m)

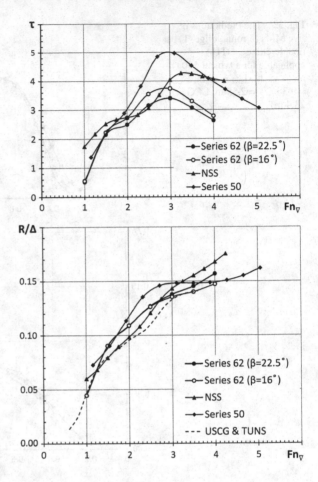

Discussion on Figs. 3.7 and 3.8

- In spite of different hull forms and displacements (500 t semi-displacement vs. 45.4 t hard chine) typical main hull form and loading parameters are almost identical in both cases, i.e. L/B = 5.6 and $L/\nabla^{1/3}$ = 6.5–6.6 (assuming L/B = $L_{wl}/B_{wl} \approx L_P/B_{PX}$ and $L/\nabla^{1/3} = L_{wl}/\nabla^{1/3} \approx L_P/\nabla^{1/3}$).
- None of the MMs for semi-displacement hull forms depend on LCG, while LCG is among the primary input variables for all hard chine forms.
- The discrepancies among comparable MMs (those in Figs. 3.7 and 3.8) are due to the fact that they actually represent different hull forms, i.e. their secondary hull form parameters are different, although their input variables are identical.

References

Almeter JM (1988) Resistance prediction and optimization of law deadrise, hard chine, stepless planing hulls. SNAME STAR symposium

Almeter JM (1993) Resistance prediction of planing hulls: state of the art. Mar Technol 30(4)

Almeter JM (2008) Avoiding common errors in high-speed craft powering predictions. In: 6th international conference on high performance Marine vehicles, Naples

Bailey D (1976) The NPL high speed round bilge displacement hull series. Maritime Technology Monograph No. 4. London : Royal institution of naval architects

Bailey D (1982) A statistical analysis of propulsion data obtained from models of high speed round bilge hulls. In: RINA symposium on small fast warships and security vessels, London

Bertram V, Mesbahi E (2004) Estimating resistance and power of fast monohulls employing artificial neural nets. International conference High Performance Marine Vehicles (HIPER), Rome

Blount DL (2014) Performance by design. ISBN 0-978-9890837-1-3

Blount DL, Fox DL (1976) Small craft power prediction. Mar Technol 13(1)

Blount DL, McGrath JA (2009) Resistance characteristics of semi-displacement mega yacht hull forms. RINA Trans, Int J Small Craft Technol 151(Part B2), July–Dec 2009

Clement PE, Blount DL (1963) Resistance tests of a systematic series of planing hull forms. SNAME Trans 71

Compton RH (1986) The resistance of a systematic series of semi-planing transom stern hulls. Marine Technol 23(4)

Couser P, Mason A, Mason G, Smith CR, Konsky BR von (2004) Artificial neural network for hull resistance prediction. In: 3rd international conference on Computer and IT Applications in the Maritime Industries (COMPIT '04), Siguenza

Davidson KSM, Suarez A (1941) Tests of twenty related models of V-bottom motor boats—U.S.E.M.B. Series 50. Report No. 170, Experimental Towing Tank, Stevens Institute of Technology, Hoboken

Delgado-Saldivar G (1993) Experimental investigation of a new series of planing hulls. M.Sc. thesis, Technical University of Nova Scotia, Halifax, Nova Scotia

De Luca F, Pensa C (2017) The Naples warped hard chine hulls systematic series. Ocean Eng 139

De Marco A, Mancini S, Miranda S, Scognamiglio R (2017) Experimental and numerical hydrodynamic analysis of a stepped planing hull. Appl Ocean Res 64

Doust DJ (1960) Statistical analysis of resistance data for trawlers. Fishing Boats of the World: 2 Fishing News (Books) Ltd., London

Farlie-Clarke AC (1975) Regression analysis of ship data. Int Shipbuilding Prog 22(251)

Fung SC (1991) Resistance and powering prediction for transom stern hull forms during early stage ship design. SNAME Trans 99

Fung SC, Leibman L (1993) Statistically-based speed-dependent powering predictions for high-speed transom stern hull forms. Chesapeake Section of SNAME

Gamulin A (1996) A semidisplacement series of ships. Int Shipbuilding Prog 43(43)

Grigoropoulos GJ, Damala DP (2001) The effect of trim on the resistance of high-speed craft, 2nd edn. In: International EURO conference on High-Performance Marine Vehicles, HIPER '01, Hamburg

Grigoropoulos GJ, Loukakis TA (1999) Resistance of double-chine large high-speed craft. Aeronautique ATMA, vol 99. Paris

Grubišić I, Begović E (2000) Resistance prediction of the fast round-bilge hulls at the concept design level. In: Proceedings of the 9th international congress of the International Association of Mediterranean, IMAM, Ischia

Hadler JB, Hubble EN, Holling HD (1974) Resistance characteristics of a systematic series of planing hull forms—Series 65. Chesapeake Section of SNAME

Holling HD, Hubble EN (1974) Model resistance data of a Series 65 hull forms applicable to hydrofoils and planing craft. NSRDC Report 4121

Holtrop J, Mennen GGJ (1982) An approximate power prediction method. Int Shipbuilding Prog 29(335)

Hubble EN (1974) Resistance of hard-chine stepless planing craft with systematic variation of hull form, longitudinal centre of gravity and loading. DTNSRDC R&D Report 4307

Jin P, Su B, Tan Z (1980) A parametric study on high-speed round bilge displacement hulls. High-Speed Surface Craft

Keuning JA, Geritsma J (1982) Resistance tests of a series of planing hull forms with 25 degrees deadrise angle. Int Shipbuilding Prog 29(337)

Keuning JA, Hillege L (2017a) The results of Delft systematic deadrise series. In: Proceedings of 14th international conference on Fast Sea Transportation (FAST 2017), Nantes

Keuning JA, Hillege L (2017b) Influence of rocker and twist and the results of the Delft systematic deadrise series. High Speed Marine Vehicles Conference (HSMV 2017), Naples

Keuning JA, Gerritsma J, Terwisga PF (1993) Resistance Tests of a series planing hull forma with 30° deadrise angle, and a calculation model based on this and similar systematic series. Int Shipbuilding Prog 40(424)

Kowalyshyn DH, Metcalf B (2006) A USCG systematic series of high speed planing hulls. SNAME Trans 114

Lahtiharju E, Karppinen T, Hellevaara M, Aitta T (1991) Resistance and seakeeping characteristics of fast transom stern hulls with systematically varied form. SNAME Trans 99

MacPherson DM (2003) Comments on reliable prediction accuracy. A HydroComp Technical Report 103

Mason A, Couser P, Mason G, Smith CR, von Konsky BR (2005) Optimisation of vessel resistance using genetic algorithms and artificial neural networks. In: 4th international conference on Computer and IT Applications in the Maritime Industries (COMPIT '05), Hamburg

Mercier JA, Savitsky D (1973) Resistance of transom-stern craft in the pre-planing regime. Davidson Laboratory Report 1667

Molland AF, Turnock SR, Hudson DA (2011) Ship resistance and propulsion—practical estimation of ship propulsive power. Cambridge University Press, ISBN 978-0-521-76052-2

Moore WL, Hawkins F (1969) Planing boat scale effects on trim and drag (TMB Series 56). NSRDC Technical Note No. 128, Washington

Morabito MG (2013) Re-analysis of Series 50 Tests of V-bottom motor boats. SNAME Trans 121

Morabito M, Snodgrass J (2012) The use of small model testing and full scale trials in the design of motor yacht. In: SNAME's 3rd chesapeake power boat symposium, Annapolis

Müller-Graf B (1999) Widerstand und hydrodynamische Eigenschaftender schnellen Knickspant-Katamarane der VWS Serie '89 (Resistance and hydrodynamic characteristics of the VWS fast hard chine catamaran Series '89). In: 20th symposium Yachtenwurf und Yachtbau, Hamburg

Radojčić D (1984) A statistical method for calculation of resistance of the stepless planing hulls. Int Shipbuilding Prog 31(364)

Radojčić D (1985) An Approximate method for calculation of resistance and trim of the planing hulls. University of Southampton, Ship Science Report No. 23. Paper presented on SNAME symposium on powerboats

Radojčić D (1991) An engineering approach to predicting the hydrodynamic performance of planing craft using computer techniques. RINA Trans 133

Radojčić D, Kalajdžić M (2018) Resistance and trim modeling of naples hard chine systematic series. RINA Trans Int J Small Craft Technol. doi:https://doi.org/10.3940/rina.ijsct.2018.b1.211

Radojčić D, Rodić T, Kostić N (1997) Resistance and trim predictions for the NPL high speed round bilge displacement hull series. RINA conference on power, performance and operability of small craft, Southampton

Radojčić D, Prinčevac M, Rodić T (1999) Resistance and trim predictions for the SKLAD semi displacement hull series. Oceanic Eng Int 3(1)

Radojčić D, Grigoropoulos GJ, Rodić T, Kuvelić T, Damala DP (2001) The resistance and trim of semi-displacement, double-chine, transom-stern hull series. In: Proceedings of 6th international conference on Fast Sea Transportation (FAST 2001), Southampton

Radojčić D, Zgradić A, Kalajdžić M, Simić A (2014a) Resistance prediction for hard chine hulls in the pre-planing regime. Pol Mar Res 21(2(82))

Radojčić D, Morabito M, Simić A, Zgradić A (2014b) Modeling with regression analysis and artificial neural networks the resistance and trim of Series 50 experiments with V-Bottom motor boats. J Ship Prod Des 30(4)

Radojčić DV, Zgradić AB, Kalajdžić MD, Simić AP (2017a) Resistance and trim modeling of systematic planing hull Series 62 (with 12.5, 25 and 30 degrees Deadrise Angles) using artificial neural networks, Part 1: the database. J Ship Prod Des 33(3)

Radojčić DV, Kalajdžić MD, Zgradić AB, Simić AP (2017b) Resistance and trim modeling of systematic planing hull Series 62 (with 12.5, 25 and 30 degrees Deadrise Angles) using artificial neural networks, Part 2: mathematical models. J Ship Prod Des 33(4)

Robinson JL (1999) Performance prediction of chine and round bilge hull forms. RINA international conference on Hydrodynamics of High Speed Craft, London

Sabit AS (1971) Regression analysis of the resistance results of the BSRA series. Int Shipbuilding Prog 18(197)

Sahoo P, Peng H, Won J, Sangarasigamany D (2011) Re-evaluation of resistance prediction for high-speed round bilge hull forms. In: Proceedings of 11th international conference on Fast Sea Transportation (FAST 2011), Honolulu

Savitsky D (1964) Hydrodynamic design of planing hulls. Mar Technol 1(1)

Savitsky D (2012) The effect of bottom warp on the performance of planing hulls. In: SNAME's 3rd Chesapeake Power Boat symposium, Annapolis

Savitsky D, Brown PW (1976) Procedure for hydrodynamic evaluation of planing hulls in smooth and rough water. Mar Technol 13(4)

Savitsky D, Roper JK, Benen L (1972) Hydrodynamic development of a high speed planing hull for rough water. In: 9th symposium Naval Hydrodynamics, ONR, Paris

Swift PM, Nowacki H, Fischer JP (1973) Estimation of Great Lakes bulk carrier resistance based on model test data regression. Mar Technol 10(4)

Tanaka H, Nakato M, Nakatake K, Ueda T, Araki S (1991) Cooperative resistance tests with geosim models of a high-speed semi-displacement craft. J SNAJ 169

Taunton DJ, Hudson DA, Shenoi RA (2010) Characteristics of a series of high speed hard chine planing hulls—Part 1: performance in calm water. RINA Trans 152, Part B2. Int J Small Craft Technol

van Hees MT (2017) Statistical and theoretical prediction methods. Encyclopedia of Maritime and Offshore Engineering, Wiley

van Oortmerssen G (1971) A power prediction method and its application to small ships. Int Shipbuilding Prog 18

van Oossanen P (1980) Resistance prediction of small high-speed displacement vessels: state of the art. Int Shipbuilding Prog 279(313)

Yegorov IT, Bunkov MM, Sadovnikov YM (1978) Propulsive performance and seaworthiness of planing vessels. Sudostroenie, leningrad (in Russian)

Chapter 4
Propeller's Open-Water Efficiency Prediction

4.1 An Overview of Modeling Propeller's Hydrodynamic Characteristics

Modeling propeller's open water hydrodynamic characteristics is in many respects different from modeling resistance, although the same tools and methods are used. Two main differences should be emphasized:

1. Dependent variables that should be modeled simultaneously are thrust coefficient (K_T) and torque coefficient (K_Q). By definition, these coefficients are interrelated (linked) through the expression for the open water efficiency—$\eta_O = (K_T/K_Q) \cdot (J/2\pi)$. There is nothing similar for modeling resistance as there is no explicit relationship between resistance and any other dependent variable. Resistance and trim, for instance, are correlated, but are not linked.
2. While the dependent variables are always K_T and K_Q, the independent ones are some or all of the following: advance coefficient ($J = v_a/nD$), pitch ratio (P/D), area ratio (A_E/A_O or A_D/A_O), number of blades (z), and cavitation number (σ or $\sigma_{0.7R}$). This pre-determination makes modeling easier, since there is no need to search for optimum independent variables best suited for a particular propeller series.

Note that although K_T-K_Q-η_O interdependence is a fact, some MMs improperly ignore this, often resulting in a chaotic η_O curve. Namely, modeling a propeller's open water hydrodynamic characteristics is, in mathematical terms, *a multiple objective (or multicriteria) optimization with constraints*. *Multiple objective* because a MM should be simultaneously obtained for both K_T and K_Q, and *with constraints* because K_T and K_Q are linked through an equality constraint $\eta_O = (K_T/K_Q) \cdot (J/2\pi)$.

The original version of this chapter was revised: For detailed information please see correction. The correction to this chapter is available at https://doi.org/10.1007/978-3-319-94899-7_8

Input variables depend on the size of the propeller series that is modeled. From the complexity viewpoint, MMs may be classified as follows (similarly done in Radojčić 1988):

Rank-1—A single propeller may be modeled satisfactorily by a third degree polynomial—K_T, $K_Q = f(J)$.
Rank-2—A small series where P/D is varied ($A_E/A_O = const.$, $z = const.$) polynomial eq. is K_T, $K_Q = f(J, P/D)$.
Rank-3—A series with only $z = const.$, polynomial eq. is K_T, $K_Q = f(J, P/D, A_E/A_O)$.
Rank-4—A series where all four variables are varied, polynomial eq. is K_T, $K_Q = f(J, P/D, A_E/A_O, z)$.

When cavitation becomes influential the cavitation number σ or $\sigma_{0.7R}$ should also be one of the independent variables (and the database comes from the cavitation tunnel experiments):

Rank-5—Extension to Rank-3 above as cavitation number is varied, so eq. is K_T, $K_Q = f(J, P/D, A_E/A_O, \sigma)$.
Rank-6—A series where all five variables are varied, polynomial eq. is K_T, $K_Q = f(J, P/D, A_E/A_O, z, \sigma)$.

Two facts should be noted here:

- All lower level MMs, belonging to Ranks-1 to 5, may be regarded as a special case of a Rank-6 MM, and
- Given that HSC propellers inherently cavitate, to a smaller or greater extent, MMs of interest in this work belong to Rank 5 or 6. However, due to a variety of reasons, models belonging to Rank 5 were developed only recently and to the best of author's knowledge Rank 6 models do not yet exist.[1]

The modeling methods discussed above imply that some mathematical expression adequately represents the dataset. A different approach is given in Bukarica (2014) where the entire dataset is a part of a MM, while the modeling of multidimensional surfaces (i.e. K_T and K_Q) is done with the *interpolating spline functions*. Although the multidimensional surface obtained through the application of splines is smooth, this approach does not seem to be of practical use, hence will not be elaborated further.

---

[1] With the exception of cases where ANN was applied directly for obtaining K_T and K_Q (e.g. Neocleous and Schizas 2002), and not for MM development (which then can be used by other users who do not have knowledge about ANN whatsoever).

4.1.1 MARIN Propeller Series

The first Rank-2 polynomials for 4- and 5-bladed Wageningen B-series propellers were presented in van Lammeren et al. (1969). After elimination of the negligible terms these polynomials had 10 terms for both K_T and K_Q. The same paper presented Rank-3 polynomials for 4- and 5-bladed B-series propellers with 18–26 terms. Finally, the entire B-series was presented with a Rank-4 MM with 39 and 47 polynomial terms for K_T and K_Q respectively, in Oosterveld and van Oossanen (1975). Ducted propeller K_a-series in various nozzles are presented with Rank-2 polynomials and have around 10 terms for each K_T and K_Q.

Polynomial equations of Rank-4 for B-series, as well as Rank-2 for various pairs of propeller + nozzle, have been finalized and have been in worldwide use for the last four decades. Traditional charts practically do not exist anymore. MARIN's book, published on the 60th anniversary of the Institute (Kuiper 1992), is the crossover between chart-based and computer-based presentation of the abovementioned Wageningen propeller series. Nowadays the purpose of charts, if provided, is only to prove the modeling was done correctly.

Note that the B-series is by far the largest series, consisting of 120 propellers that were tested over a period of almost 40 years. Experiments were conducted in different facilities, under different conditions and for different Reynolds numbers, so that regression analysis, amongst other faired experimental results and formed unified database too. Derived polynomials are valid for $Rn = 2 \times 10^6$; corrections for other Reynolds numbers are given with the additional polynomials ΔK_T and ΔK_Q. Note that the traditional charts did not even state the Rn for which they were valid.

The B-series, with inner blade sections of airfoil type and outer sections of segmental type, is a fixed-pitch general purpose propeller series, and is currently used extensively for both design and benchmarking purposes (see Carlton 2012). B-series polynomials are usable for speeds of up to 20–30 kn or so, for $z = 2$–7, $P/D = 0.5$–1.4, and $A_E/A_O = 0.3$–1.05. Note however that caution should be exercised with the boundaries of applicability (see Radojčić 1985).

Common open-water K_T-K_Q-J charts are practically just graphical presentation of B-series polynomials. However, there are other representations or formats (Troost B_P-δ and B_u-δ for instance) that enable easier propeller optimization, and are important mostly if charts, rather than MMs, are used. All parameters necessary for other formats may be easily recalculated from the open water K_T-K_Q-J values. Consequently, there are several propeller optimization tools based on the B-series polynomials, often developed with chart presentation in mind (see for instance Radojčić 1985; Yosifov et al. 1986; Loukakis and Gelegeris 1989; Shen and Marchal 1993 etc.). Application of an ANN for selection of a maximum efficiency ship propeller is given in Matulja et al. (2010), for instance.

Propeller characteristics used for ship maneuvering (stopping, astern sailing etc.) are usually presented with four quadrant diagrams where $C_T{}^*$-$C_Q{}^*$-β are used instead of K_T-K_Q-J parameters. $C_T{}^*$ and $C_Q{}^*$ are, respectively, thrust and torque indexes, while β is a hydrodynamic pitch angle at 0.7R. The effect of P/D, A_E/A_O, and z in the 4-quadrant presentation is analyzed traditionally with a Fourier analysis (e.g. van Lammeren et al. 1969), but the ANN technique was recently also used (see Roddy et al. 2006). The B-series and ducted K_a-series propellers in nozzles, both being trademarks of MARIN, are succeeded by new 4- and 5-bladed C- and ducted D-Series of Controllable Pitch Propellers (CPP), as described in Dang et al. (2013) where 2-quadrant diagrams in C_T^*-C_Q^*-β format are given (important for propeller's off-design conditions).

Note that a direct relationship between open water K_T, K_Q, and J, and respectively, 4-quadrant C_T^*, C_Q^*, and β parameters does exist. Conventional open water K_T-K_Q-J presentation, corresponding to the first quadrant, is in fact just a special case of the 4-quadrant presentation. Being simpler and more practical for the subject in hand (i.e. power prediction), traditional K_T-K_Q-J coefficients are used in all further discussions here.

When RPM is required K_T-K_Q-J can be easily preformatted to K_T/J^2 or K_Q/J^3, while the known input values are R, D, and v, or P, D, and v, respectively. However, if D is required and RPM is known, transformation K_T/J^4 or K_Q/J^5, respectively would be adequate. The thrust-loading format η_O, $J = f(K_T/J^2)$ is frequently used by the HSC community.

Summarizing, it may be concluded that the propeller series, compared to ship systematic series, are more resistant to the passage of time (van Hees 2017). The reason for this is the fact that propellers' hydrodynamic characteristics can be modeled with just a few parameters, which is not the case with the ship hull hydrodynamic characteristics.

4.2 Mathematical Modeling of K_T, K_Q, and η_O of High-Speed Propellers

While MMs for HSC resistance prediction were presented in a chronological order, with each more or less more sophisticated than the previous one, MMs for propellers, ranked as explained above, are presented from the application viewpoint, i.e. going from (for the HSC) lower to higher vessel speeds.

4.2.1 AEW and KCA Propeller Series

HSC propellers should be more resistant to cavitation than the B-series. Flat-faced, segmental section propellers are inherently more resistant to the inception of cavitation than those with the airfoil sections. They also have good open-water characteristics and are easier to manufacture and repair. Amongst the first systematic tests of these propellers were those reported by Gawn (1953) and Gawn and Burrill (1957). Parameter space of the series is roughly P/D = 0.6–2.0, A_E/A_O = 0.5–1.1, and σ =

0.5–6.3, but the range of MMs discussed here vary from one MM to another. Both series consist of similar three bladed propellers, but are tested in different experimental facilities, and under different conditions. The Gawn (1953) AEW data are valid for open-water conditions, and Gawn and Burrill (1957) KCA data is for atmospheric and cavitating conditions. Note that the results of open-water experiments and cavitation experiments at atmospheric conditions are not necessarily the same, as is often assumed, but any further discussion on this topic is beyond the scope of this work.

Based on the aforementioned experimental data, four MM were developed. The first model is based on the AEW data (for non-cavitating conditions only), and the other three on the KCA data. Note that up to 10% back cavitation might be regarded as non-cavitating, as the breakdown point is not reached yet. The first three MMs were developed using multiple regression techniques, while the fourth one is derived by ANN. These MMs are compared in Radojčić et al. (2009).

Model 1—(Blount and Hubble 1981)

Consists of separate equations for the non-cavitating (open-water) and for the cavitating regime. Open water equations are actually the B series polynomials with 39 and 47 terms for K_T and K_Q respectively (i.e. Rank-4 MM), but new regression coefficients were evaluated. The cavitating regime however, is represented with a relatively simple set of equations (see Blount and Fox 1978, also discussed in Sect. 4.3). Note that from the mathematical viewpoint, use of the same polynomial terms to model different test data, i.e. AEW and B-series propellers, and hence extending model's validity to four bladed propellers, is not correct, although it may be considered practical.

Model 2—(Kozhukharov 1986)

Consists of a pair of single equations for both regimes, with 121 and 116 polynomial terms for K_T and K_Q respectively. Hence this is a Rank-5 MM, although a relatively simple function transformation was used for the dependent variables.

Model 3—(Radojčić 1988)

Consists of a pair of equations for the non-cavitating, and an additional pair for the cavitating regime. The first set, belonging to Rank-3 MMs, has 16 and 17 polynomial terms for K_T and K_Q. The second set has 20 and 18 terms for evaluation of ΔK_T and ΔK_Q reductions for cavitating conditions. These are f(A_D/A_O, P/D, $\sigma_{0.7R}$, K_T) so that for the cavitating regime there are in total 36 and 35 terms respectively.

Model 4— (Koushan 2007)

Separate equations are derived for the non-cavitating and cavitating conditions, with 34 equation terms for each K_T and K_Q for the non-cavitating regime, and 89 and 107 terms respectively for the cavitating regime. Linear and hyperbolic tangent functions were used for the activation function for the non-cavitating and the cavitating regime, respectively. In a broader sense the first set of equations would belong to the Rank-3 MM and the second to the Rank-5 MM.

Discussion of Models 1 to 4—(Radojčić et al. 2009)

For all four abovementioned MMs additional narrower, and more conservative, boundaries of applicability are suggested in Radojčić et al. (2009). Also, special attention was paid to the η_O representation for all four models, since modeling errors in independently evaluated K_T and K_Q are well illustrated by η_o (see Radojčić 1988). Model 1 seems to have overlooked this in the cavitating regime only. Model 2 and Model 4 produce inconsistent values of η_O even for non-cavitating regimes. Models 1 and 3 however, do have disadvantages around the K_T and K_Q cavitation breakdown points, as the transition zone (multidimensional plane) should be faired smoothly into the open-water data (another multidimensional plane).

Model 1 and Model 3 are valid for the non-cavitating regime all the way through to the inception of cavitation (intersection of open-water and transition zone, i.e. K_T and K_Q breakdown points). Model 4 however, consists of separate equations for the cavitating and non-cavitating regimes for the whole J-range. Note that from the physical viewpoint this is not correct. Model 2 is based on the single equations for both cavitating and non-cavitating regimes. In principle this is correct, but is difficult for modeling because the character of K_T and K_Q curves are completely different for the non-cavitating and cavitating conditions.

Finally, it was concluded that the advantage of Model 1 is its simplicity and validity for heavily-cavitating propellers. Model 3 is probably the best for non-cavitating conditions, while Model 4 is advantageous for the transition zone. Model 2 appears to have no advantages compared to the others. Model 1 is applicable for 3- and 4-bladed propellers, while all other models are valid for 3-bladed propellers only (as is the case for both AEW and KCA propeller series).

4.2.2 Newton-Rader Propeller Series

Newton-Rader three bladed propeller series (Newton and Rader 1961) covered P/D, A_E/A_O, and σ range from about 1–2, 0.5–1.1, and 0.25–2.5, respectively. This hollow-faced blade section series is intended for very high speeds of 50 or so knots, while the flat-faced segmental section propellers are intended for speeds of up to approximately 40 kn. Note however, that this kind of propellers generally require custom-design and manufacture, in contrast to the B-series and flat-faced segmental section propellers.

There are two MMs that represent the Newton-Rader propeller series. The first one (Kozhukarov and Zlatev 1983) was developed using the multiple regression technique, while the second one (Koushan 2005) used ANN. Both MMs are very similar to Model 2 and Model 4 respectively, for the segmental section propeller series. The regression Rank 5 model has 101 polynomial terms for both K_T and K_Q, while ANN derived models for the non-cavitating and cavitating conditions have, respectively, 34 and 101 equation terms for each K_T and K_Q. Note however, that the author could not obtain reasonable results using K_T and K_Q polynomials given in Kozhukarov and Zlatev (1983). This was also reported by Diadola and Johnson (1993)

Taking into account: (a) the abovementioned similarity between the MMs (Kozhukharov and Zlatev 1983 vs. Kozhukharov 1986, and Koushan 2005 vs. 2007), and (b) negligence of comparison of recently developed ANN models with the existing regression model (Koushan 2005, and Kozhukarov and Zlatev 1983), it might be expected that most of the conclusions derived in Radojčić et al. (2009) for AEW and KCA propeller series are also valid for the Newton-Rader MMs.

4.2.3 Swedish SSPA Ma and Russian SK Series

Both the Swedish SSPA Ma 3- and 5-bladed series ($A_E/A_O = 0.75$–1.2, $P/D = 1$–1.45, $\sigma = 0.25$–atm., see Lindgren 1961), and Russian 3-bladed SK series ($A_E/A_O = 0.65$–1.1, $P/D = 1$–1.8, $\sigma = 0.3$–atm., see Mavludov et al. 1982) are similar to the Newton-Rader series.

For the SSPA Ma 3-075 propeller series for open water conditions the polynomial equations for evaluation of K_T and $K_Q = f(J, \sigma, P/D)$, with 22 and 18 terms respectively, are given in Blount and Bjarne (1989). Note that K_T and $K_Q = f(\sigma)$, which is unusual.

Concerning the SK series, the author initiated a project (Milićević 1998) with the purpose to regress the K_T and K_Q data for cavitating conditions, as done in Radojčić (1988), but without success. Namely, K_T, K_Q, and η_O curves, as published in Mavludov et al. (1982), were found not to be consistent for the SK series (K_T and K_Q curves are linked through $\eta_O = (K_T/K_Q) \cdot (J/2\pi)$ as already discussed in Sect. 4.1). Only a Rank 3 MM for open water conditions with 21 and 26 polynomial terms for K_T and K_Q, respectively, could be obtained.

4.2.4 SPP Series

Surface Piercing Propellers (SPP) are also intended for very high speeds—above 40 or so knots (see Allison 1978 and Kruppa 1990). SPP's modus operandi at high speeds is: (a) reduced cavitation danger due to ventilation on the suction side (the pressure side remains fully wetted), and (b) avoidance of appendage resistance (the appendages are above the water level). Additional side benefits include a vertical force, but only when the SPPs are properly installed. SPPs have few disadvantages; torque absorption depends on propeller immersion which, depending on the SPP type, may be influenced by shaft inclination. Consequently, amongst the SPP's input variables is immersion ratio (h/D), or shaft inclination (Ψ).

A MM for the SPP series based on five 4-bladed Rolla Series ($A_E/A_O = 0.8$, $P/D = 0.9$–1.6) is given in Radojčić and Matić (1997). Tests were carried out in a free-surface cavitation tunnel under $\sigma = 0.2$, 0.5, and atmospheric pressure, while the immersion ratio (h/D) was 30, 47.4, and 58% corresponding to shaft inclination of 4°, 8°, and 12°, respectively (see Rose and Kruppa 1991; Rose et al. 1993). Consequently, the

SPP MM consists of three *cavitation independent* pairs, each with 13–20 polynomial terms of K_T, $K_Q = f$ (h/D, P/D, J), one for each of the tested cavitation numbers σ. Special attention was paid to selecting an optimum K_T and K_Q pair to give the best representation of $\eta_O = (K_T/K_Q) \cdot (J/2\pi)$. This however meant that, from the statistical point of view, the finally selected individual K_T and K_Q curves were not the best fit.

Experimental results of another 4 and 5 bladed systematic SPP series with P/D ratio between 0.8 and 1.4 are represented by regression equations for $K_T{'}$ and $K_Q{'}$ as a f(J_ψ, P/D), see Ferrando et al. (2007). That is, the main parameters are slightly modified, so that the advance, thrust and torque coefficients are revised to $J_\psi = v_A \cdot \cos\Psi/nD$, $K_T{'} = T/(\rho n^2 D^2 A_O{'})$, and $K_Q{'} = Q/(\rho n^2 D^3 A_O{'})$ respectively, where Ψ is longitudinal shaft inclination and $A_O{'}$ is immersed propeller area. As a result of this unique *SPP nondimensionalization*, different K_T and K_Q curves for various shaft inclinations respectively, collapse into single curves $K_T{'}$ and $K_Q{'}$, but only for advance coefficients J_ψ above the critical values J_{CR}. This fact enabled formulation of the relatively simple equations $K_T{'}$, $K_Q{'} = f(J_\psi, P/D)$, valid for $J_\psi > J_{CR}$, which actually correspond to the partially vented regime. Note that the fully vented regime (i.e. regime below J_{CR}) is not modeled.

4.3 Loading Criteria for High-Speed Propellers

The design objectives for high-speed propellers include maximal efficiency, minimal cavitation, and minimal propeller diameter. Some guidelines for allowable propeller loading are required to achieve these objectives, particularly when cavitation becomes significant. The 10% back cavitation curve (stemming from the segmental section Gawn-Burrill KCA propeller series) is usually assumed to be an erosion-free criterion for high-speed propellers. Note that the loss of thrust due to excessive cavitation (i.e. the thrust breakdown point) usually occurs between 10 and 20% back cavitation.

Further development of loading limits requires a new propeller format based on the local cavitation number ($\sigma_{0.7R}$, using resultant water velocity at 0.7 radius—$v_{0.7R}$), thrust and torque load coefficients τ_C and Q_C, respectively. These are:

$$\sigma_{0.7R} = \left(p_A + p_H - p_V\right)/\left(\rho/2 \cdot v_{0.7R}{}^2\right)$$
$$\tau_C = T/\left(\rho/2 \cdot A_P \cdot v_{0.7R}{}^2\right)$$
$$Q_C = Q/\left(\rho/2 \cdot D \cdot A_P \cdot v_{0.7R}{}^2\right)$$

For each blade section, three groups of simple equations (in the log-log coordinate system these are straight lines) approximate the data scatter τ_C & Q_C versus $\sigma_{0.7R}$:

1. τ_C & Q_C showing inception of 10% back cavitation,
2. τ_C & Q_C indicating maximal K_T & K_Q values respectively, in the transition (partly cavitating) zone, and
3. τ_C & Q_C in the fully cavitating zone.

Essentially, the B-, KCA-, Newton-Rader-, and super-cavitating propeller series (surface piercing propeller (SPP) series was added later) characterize the various blade sections, each represented with a distinct set of equations. These equations actually outline a high-speed propeller loading criteria, i.e. MMs ready for computerization. The guidelines (obtained under experimental conditions) suggest the upper limits of τ_C and Q_C for transition and fully cavitating regimes, with about 90% confidence. For the full-scale conditions, it is recommended to reduce the limits to 80% of maximal allowable τ_C & Q_C value. The viability and usefulness of these simple loading criteria has been proven through experience since their initial release in 1978 (Blount and Fox 1978). The complete set of equations, including those for SPPs, is given in Blount's book (2014). See also the recommendations given in Bjarne (1993).

Note that these design criteria are by no means sufficient metrics for sizing an optimal propeller. That is, loading limits can easily identify an optimal, but inappropriate propeller with, for instance, a large pitch ratio (e.g. P/D ≈ 2). The off-design conditions must also be checked, resulting in overall optimal propeller characteristics (typically with P/D between 1.1 and 1.4, see Blount and Fox 1978) at the cost of slight efficiency loss at one (usually maximal) speed. Note that these guidelines should be used with care since the abovementioned criteria are functions only of propeller characteristics, while other influential parameters are omitted (e.g. wake).

4.4 Recommended Mathematical Models for High-Speed Propellers

The following MMs for HSC propellers are recommended:

1. B series (3–6 blades, with inner blade sections of aerofoil type, outer of segmental type, for non-cavitating regime and speeds of up to 20–30 kn),
2. AEW/KCA series (3- and 4-bladed, segmental flat-faced sections throughout, non-cavitating and cavitating regimes, for speeds 25–40+ kn), and
3. Newton-Rader series (3-bladed, hollow-faced blade sections, cavitating regime, for speeds 40–50+ kn).

For speeds above 30, 40, and 50 knots, the waterjets, surface piercing (SPP), and submerged fully cavitating propellers respectively, should be considered (in-depth discussion is given in Blount 2014).

Commercial off-the-stock propellers usually belong to the first two groups, while the hollow-faced blade section propellers are usually custom-made. In some cases flat-faced propellers are convenient for cupping (bending of propeller's trailing edge, corresponding to pitch increase), which bring flat-faced sections closer to hollow-faced sections, hence increase their ability in partly- and fully-cavitating regimes (see Blount and Hubble 1981; MacPherson 1997, for instance).

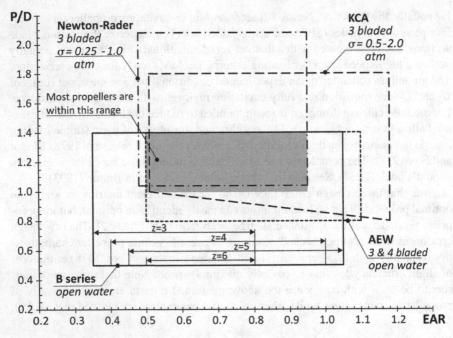

Fig. 4.1 Applicability range of MMs for B-, AEW-, KCA-, and Newton-Rader Series (Oosterveld and van Oossanen 1975; Blount and Hubble 1981; Radojčić 1988; Koushan 2007; 2005 respectively)

Propellers of the second group are of principal interest for HSC. Recommended MMs for segmental flat-faced section propellers are:

- 4-bladed, non-cavitating regime—Blount and Hubble (1981)
- 3-bladed, non-cavitating regime—Radojčić (1988)
- 3-bladed, transition region—Koushan (2007)
- 3-bladed, cavitating regime—Blount and Hubble (1981), Radojčić (1988) and Koushan (2007), but use all three with caution.

For quick reference the recommended MMs for B, AEW/KCA, and the Newton-Rader series are presented in Table 4.1. Their boundaries of applicability are shown in Fig. 4.1. Other MMs mentioned earlier are incomplete (with respect to the range of applicability), do not seem to have any advantage over the recommended ones, or are not considered to be reliable enough. Note that the loading criteria equations are partly incorporated in the MM for the cavitating regime of Blount and Hubble (1981), shown in Table 4.1.

Table 4.1 Recommended MMs for high-speed propellers with their boundaries of applicability

Name Reference	Mathematical model	Boundaries of applicability	Remarks
B series (see 4.1.1) Oosterveld and van Oossanen (1975)	For non-cavitating conditions (Based on B series data) $$K_T = \sum_{i=1}^{39} (C_{Ti} \cdot (J)^{s_i} \cdot (P/D)^{t_i} \cdot (EAR)^{u_i} \cdot (z)^{v_i})$$ $$K_Q = \sum_{i=1}^{47} (C_{Qi} \cdot (J)^{s_i} \cdot (P/D)^{t_i} \cdot (EAR)^{u_i} \cdot (z)^{v_i})$$ C_T, C_Q, s, t, u, v—coefficients for non-cavitating conditions Valid for $R_n = 2 \times 10^6$, for other R_n corrections ΔK_T and ΔK_Q	For non-cavitating conditions z = 3–6 $0.5 < P/D < 1.4$ $0.35 < EAR < 1.05$ Extreme values should be avoided	Advantages: By far the largest and best known fixed-pitch, general purpose series. Used for benchmarking purposes. Disadvantages: For speeds up to 20–30 kn (inner blade sections aerofoil).
AEW series (see 4.2.1, Model 1, and Sect. 4.3) Blount and Hubble (1981)	For non-cavitating conditions (Based on AEW data) $$K_T = \sum_{i=1}^{39} (C_{Ti} \cdot (J)^{s_i} \cdot (P/D)^{t_i} \cdot (EAR)^{u_i} \cdot (z)^{v_i})$$ $$K_Q = \sum_{i=1}^{47} (C_{Qi} \cdot (J)^{s_i} \cdot (P/D)^{t_i} \cdot (EAR)^{u_i} \cdot (z)^{v_i})$$ C_T, C_Q, s, t, u, v—coefficients for non-cavitating conditions For cavitating conditions (Based on Blount and Fox 1978) $$K_T = 0.393 \cdot \tau_c \cdot EAR \cdot (1.067 - 0.229 \cdot P/D) \cdot (J^2 + 4.84)$$ $$K_Q = 0.393 \cdot Q_c \cdot EAR \cdot (1.067 - 0.229 \cdot P/D) \cdot (J^2 + 4.84)$$ For transition region: $$\tau_c = 1.2 \cdot \sigma_{0.7R}$$ $$Q_c = 0.2 \cdot P/D \cdot \sigma_{0.7R}^{(0.7+0.31 \cdot EAR^{0.9})}$$ For fully developed cavitation: $$\tau_{cx} = 0.0725 \cdot P/D - 0.034 \cdot EAR$$ $$Q_{cx} = (0.0185 \cdot (P/D)^2 - 0.0166 \cdot P/D + 0.00594/\sqrt{EAR}$$	For non-cavitating conditions z = 3 and 4 $0.80 \le P/D \le 1.40$ $0.50 \le EAR \le 1.10$ For cavitating conditions $0.60 \le P/D \le 2.00$ $\tau_{cx} \le \tau_c$ $Q_{cx} \le Q_c$ $\sigma_{0.7R} \le 0.4$	Advantages: Applicable for 3- and 4-bladed segmental section propellers. MM valid for heavily cavitating conditions. Disadvantages: Transition zone, non-cav. to cav. not smooth. $\eta_O = f(K_T, K_Q)$ not good for cavitating regime.

(continued)

Table 4.1 (continued)

Name Reference	Mathematical model	Boundaries of applicability	Remarks
KCA series (see 4.2.1, Model 3) Radojčić (1988)	For non-cavitating conditions (Based on KCA data) $K_T = \sum_1^{16} C_T \cdot 10^e \cdot (DAR)^x \cdot (P/D)^y \cdot (J)^z$ $K_Q = \sum_1^{17} (C_Q \cdot 10^e \cdot (DAR)^x \cdot (P/D)^y \cdot (J)^z)$ C_T, C_Q, e, x, y, z—coefficients for non-cavitating conditions Additional equations for cavitating conditions (Based on KCA data) $\Delta K_T = \sum_1^{20} (d_T \cdot 10^e \cdot (DAR)^s \cdot (P/D)^v \cdot (\sigma_{0.7R})^t \cdot (K_T)^u)$ $\Delta K_Q = \sum_1^{20} (d_Q \cdot 10^e \cdot (DAR)^s \cdot (P/D)^v \cdot (\sigma_{0.7R})^t \cdot (K_T)^u)$ d_T, d_Q, e, s, v, t, u—coefficients for cavitating conditions	For non-cavitating conditions $0.5 \le DAR \le 1.1$ $0.8 \le P/D \le 2.0$ $J \ge 0.3$ $J_{min} = 0.492 \cdot P/D + 0.031$ $K_T \le \sqrt{J/2.5} - 0.1$ For cavitating conditions for $\sigma < 1.00$ $J_{min} = 0.571 \cdot P/D + 0.021$ for $\sigma \ge 1.00$ $J_{min} = 0.492 \cdot P/D + 0.031$ $1.25 - 0.3 \cdot (DAR) - 0.2 \cdot \sigma \le P/D \le 1.8$ $K_T \le \sqrt{J/2.5} - 0.1 - (0.07/\sigma)$ $\sigma \ge 0.50$ $\Delta K_T \ge 0, \Delta K_Q \ge 0$ $\eta_o \ge 0.20, \eta_{atm} \ge \eta_{cav}$	Advantages: Good for non-cavitating 3-bladed segmental section propellers. Good representation of $\eta_O = f(K_T, K_Q)$. Disadvantages: Transition zone non-cav. to cav. not good enough.
KCA series (see 4.2.1, Model 4) Koushan (2007)	Equation for non-cavitating and cavitating conditions (Based on KCA data) $Y = \frac{f_o \left(C + \sum_{i=1}^{n} \left(O_i \cdot \tanh \left(a_i + \sum_{k=1}^{m} (h_{k,i} \cdot (D_k \cdot X_k + E_k)) \right) \right) \right) - G}{L}$ Y—K_T or K_Q coefficient a, C, D, E, G, h, L, O—constants (depend whether Y is K_T or K_Q and if cavitation is present or not) X—Input variables in following order: J, DAR, P/D, and σ 89 and 107 terms for K_T and K_Q respectively for cavitating conditions	For non-cavitating and cavitating conditions $0.60 \le P/D \le 2.00$ $0.5 \le DAR \le 1.1$ $0.50 \le \sigma \le 2.00$ $J_{min} = 0.4847 \cdot P/D + 0.0339$ $(P/D)_{min} = -0.67 + 0.83 \cdot DAR^{-0.32} + 0.61 \cdot \sigma^2 - 1.64 \cdot \sigma$	Advantages: Good for transition zone. Single function for non-cav. and cav. regimes. Disadvantages: $\eta_O = f(K_T, K_Q)$ inconsistent.

(continued)

Table 4.1 (continued)

Name Reference	Mathematical model	Boundaries of applicability	Remarks
Newton-Rader series (see 4.2.2) Koushan (2005)	Separate equations for non-cavitating and cavitating conditions (Based on Newton-Rader data) $$Y = \frac{f_o\left(C + \sum_{j=1}^{m}\left(O_j \cdot \tanh\left(a_j + \sum_{i=1}^{m}\left(h_{k,i} \cdot (D_k \cdot X_k + E_k)\right)\right)\right)\right)-G}{L}$$ Y—K_T or K_Q coefficient a, C, D, E, G, h, L, O—constants (depend whether Y is K_T or K_Q and if cavitation is present or not) X—Input variables in following order: J, EAR, P/D and σ 34 equation terms for both K_T and K_Q for non-cavitating conditions 101 equation terms for both K_T and K_Q for cavitating conditions	For non-cavitating and cavitating conditions $1.04 \leq P/D \leq 2.08$ $0.48 \leq EAR \leq 0.95$ $0.25 \leq \sigma \leq 1.00$ For non-cavitating conditions only $J_{min} = 0.538 \cdot P/D + 0.129$ For cavitating conditions only $J_{min} = \frac{P}{D}(1.08 \cdot \sigma^2 - 1.585 \cdot \sigma + 1.013)$	Advantages: The only MM for Newton-Rader propeller series. Disadvantages: $\eta_O = f(K_T, K_Q)$ inconsistent for both regimes.

4.4.1   Some Typical Examples

Comparison of MMs with ad hoc chosen series' propellers upon which those MMs are based are shown in Figs. 4.2 and 4.3.

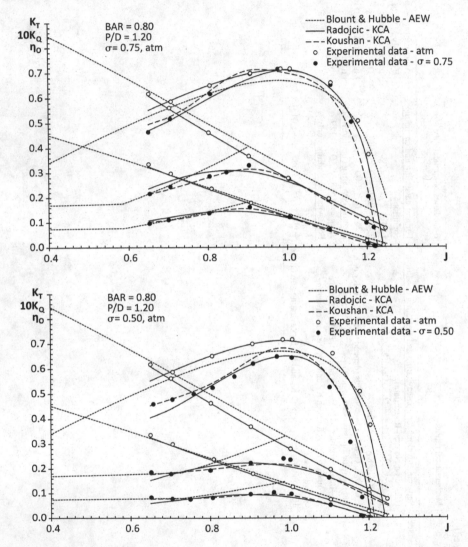

Fig. 4.2 Comparison of three MMs with KCA 112 propeller (DAR = 0.8, P/D = 1.2, σ = atm, 0.5, and 0.75)

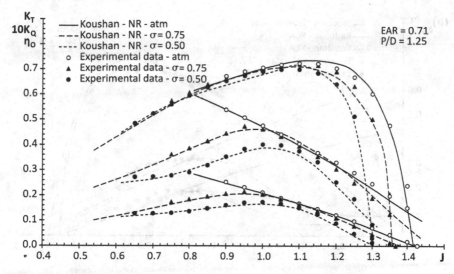

Fig. 4.3 Comparison of Koushan's (2005) MM with a parent Newton-Rader propeller A3/71/125 (EAR = 0.71, P/D = 1.25, σ = atm, 0.75, and 0.50)

A comparison with the full scale tested commercial propellers is shown in Fig. 4.4. Experimental data is from Denny et al. (1988)[2]; some propellers have subsequently been retested and were used in discussion of Radojčić (1988). Note that the MMs are based on the laboratory propellers, while the commercially available propellers are not manufactured to the same tolerances. Moreover, experiments are not always repeatable; see Fig. 4.4a.

Note: For the KCA series, DAR (Developed Blade Area) is used. All other series use EAR (Expanded Area Ratio). Consequently, BAR (Blade Area Ratio) notation is used in Figs. 4.2 and 4.4, denoting that BAR value is used as DAR and EAR values for KCA and AEW series, respectively. It is, however assumed that differences between DAR and EAR are negligible. Namely, EAR ≈ 0.34 · DAR · [2.75 + DAR/z]; see O'Brien (1969). Accordingly, for Fig. 4.1, KCA series' DAR is recalculated to EAR.

---

[2]This propeller series was modeled in Neocleous and Schizas (2002) but is not addressed here due to reasons given in Sect. 1.5.3. See also Footnote 1, Sect. 4.1.

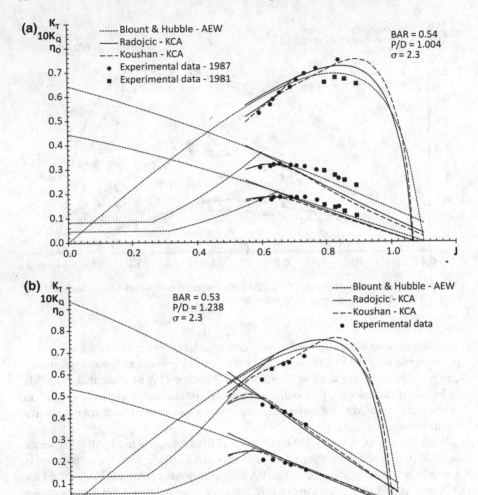

Fig. 4.4 Comparison of three MMs with commercial propellers **a** EAR = 0.54 (EAR$_{nominal}$ = 0.50), P/D = 1.004 (P/D$_{nominal}$ = 1.00), σ = 2.3, **b** EAR = 0.53 (EAR$_{nominal}$ = 0.50), P/D = 1.238 (P/D$_{nominal}$ = 1.25), σ = 2.3

References

Allison J (1978) Propellers for high performance craft. Mar Technol 15(4)

Bjarne E (1993) Completely submerged propellers for high speed craft. In: Proceedings of 2nd international conference on fast sea transportation (FAST '93), Yokohama

Blount DL (2014) Performance by design. ISBN 0-978-9890837-1-3

Blount DL, Bjarne E (1989) Design and selection of propulsors for high speed craft. In: 7th lips propeller symposium, Nordwijk-on-Sea

Blount DL, Fox DL (1978) Design considerations for propellers in cavitating environment. Mar Technol 15(2)

Blount DL, Hubble EN (1981) Sizing segmental section commercially available propellers for small craft. In: Propellers '81 symposium, SNAME, Virginia Beach

Bukarica M (2014) Mathematical modeling of propeller series. Part B2, Int J Small Craft Technol (RINA Trans) 156, July–Dec

Carlton JC (2012) Marine propellers and propulsion, 3rd edn, Butterworth-Heinemann, ISBN 9780080971230

Dang J, van den Boom HJJ, Ligtelijn JT (2013) The Wageningen C- and D-series propellers. In: Proceedings of 12th international conference on fast sea transportation (FAST 2013), Amsterdam

Denny SB, Puckette LT, Hubble EN, Smith SK, Najarian RF (1988) A new usable propeller series. SNAME, Hampton Road Section

Diadola JC, Johnson MF (1993) Software user's manual for propeller selection and optimization program (PSOP). SNAME Technical and Research Bulletin No. 7-7

Ferrando M, Crotti S, Viviani M (2007) Performance of a family of surface piercing propellers. In: 2nd International conference on marine research and transportation (ICMRT), Ischia

Gawn RWL (1953) Effect of pitch and blade width on propeller performance. INA Trans 95

Gawn RWL, Burrill LC (1957) Effect of cavitation on the performance of a series of 16 in model propellers. INA Trans 99

Koushan K (2005) Mathematical expressions of thrust and torque of Newton-Rader propeller series for high speed crafts using artificial neural networks. In: Proceedings of 8th international conference on fast sea transportation (FAST 2005), St. Petersburg

Koushan K (2007) Mathematical expressions of thrust and torque of Gawn-Burrill propeller series for high speed crafts using artificial neural networks. In: Proceedings of 9th international conference on fast sea transportation (FAST 2007), Shanghai

Kozhukarov PG (1986) Regression analysis of Gawn-Burrill series for application in computer-aided high-speed propeller design. In: Proceedings. 5th international conference on high-speed surface craft, Southampton

Kozhukarov PG, Zlatev ZZ (1983) Cavitating propeller characteristics and their use in propeller design. In: High speed surface craft conference, London

Kruppa C (1990) Propulsion systems for high speed marine vehicles. In: Second conference on high speed marine craft, Kristiansand

Kuiper G (1992) The Wageningen propeller series. MARIN Publication 92-001 (ISBN 90-900 7247-0)

Lindgren H (1961) Model tests With a family of three and five bladed propellers. SSPA Publication no 47

Loukakis TA, Gelegeris GJ (1989) A new form of optimization diagrams for preliminary propeller design. RINA Trans, Part B 131

MacPherson DM (1997) Small propeller cup: a proposed geometry standard and a new performance model. In: SNAME propellers/shafting symposium, Virginia Beach

Matulja D, Dejhalla R, Bukovac O (2010) Application of an artificial neural network to the selection of a maximum efficiency ship screw propeller. J Ship Prod Des 26(3)

Mavludov MA, Roussetsky AA, Sadovnikov YM, Fisher EA (1982) Propellers for high speed ships. Sudostroenie, Leningrad (in Russian)

Milićević M (1998) Mathematical modeling of supercavitating SK series. Diploma degree thesis, Faculty of Mechanical Engineering, Department of Naval Architecture, University of Belgrade (in Serbian)

Neocleous CC, Schizas CN (2002) Artificial neural networks in estimating marine propeller cavitation. In: Proceedings of the international joint conference on neural networks, vol 2

Newton RN, Rader HP (1961) Performance data of propellers for high-speed craft. RINA Trans 103(2)

O'Brien TP (1969) The design of marine screw propellers. Hutchinson and Co. Publishers Ltd., London

Oosterveld MWC, van Oossanen P (1975) further computer-analyzed data of the Wageningen B-screw series. Int Shipbuilding Prog 22(251)

Radojčić D (1985) Optimal preliminary propeller design using nonlinear constrained mathematical programming technique. University of Southampton, Ship Science Report no 21

Radojčić D (1988) Mathematical model of segmental section propeller series for open-water and cavitating conditions applicable in CAD. In: Propellers '88 symposium, SNAME, Virginia Beach

Radojčić D, Matić D (1997) Regression analysis of surface piercing propeller series. In: High speed marine vehicles conference (HSMV 1997), Sorrento

Radojčić D, Simić A, Kalajdžić M (2009) Fifty years of the Gawn-Burrill KCA propeller series. Part B2, Int J Small Craft Technol (RINA Trans) 151, July–Dec

Roddy RF, Hess DE, Faller W (2006) Neural network predictions of the 4-quadrant Wageningen propeller series. NSWCCD-50-TR-2006/004, DTMB Carderock Division, Bethesda

Rose J, Kruppa C (1991) Surface piercing propellers, methodical series model test results. In: Proceedings of 1st international conference on fast sea transportation (FAST '91), Trondheim

Rose J, Kruppa C, Koushan K (1993) Surface piercing propellers, propeller/hull interaction. In: Proceedings of 2nd international conference on fast sea transportation (FAST '93), Yokohama

Shen Y, Marchal LJ (1993) Expressions of the B_P-δ diagrams in polynomial for marine propeller series. In: RINA W10 (1993) paper issued for written discussion

van Hees MT (2017) Statistical and theoretical prediction methods. In: Encyclopedia of maritime and offshore engineering, Wiley

van Lammeren WPA, van Manen JD, Oosterveld MWC (1969) The Wageningen B-Screw series. SNAME Trans 77

Yosifov K, Zlatev Z, Staneva A (1986) Optimum characteristics equations for the 'K-J' propeller design charts, based on the Wageningen B-screw series. In: International shipbuilding progress, vol 33, no 382

Chapter 5
Additional Resistance Components and Propulsive Coefficients

5.1 Evaluation of In-Service Power Performance

Additional components necessary for the evaluation of a HSC's in-service power performance are briefly discussed here (see also Sect. 1.3). These components consist of those that:

1. Increase the resistance from bare hull total resistance in deep and calm water to in-service total resistance (i.e. from R_T to R_T*), and
2. Account for the hull-propeller interaction (i.e. propulsive coefficients).

Note that there are different ways how R_T* is split into its components; see for instance, Müller-Graf (1981, 1997a) and Savitsky (1981) for round bilge semi-displacement and hard chine planing hulls, respectively. For better insights of both resistance components and hull-propeller interaction, see ITTC (2011) "Recommended Procedures and Guidelines for High Speed Marine Vehicles". Practical and up-to-date recommendations for the evaluation of various additional components, as well as step-by-step HSC powering procedures, are discussed in the Blount book (2014).

5.2 Resistance Components—Calm and Deep Water

5.2.1 Appendages

In general, the approach for appendage resistance evaluation of semi-displacement (round bilge) and planing (hard chine) vessels is similar, except that the appendage configuration is slightly different (rudder type may differ, skeg may be applied to semi-displacement round bilge but not planing hulls, etc.).

© The Author(s), under exclusive licence to Springer Nature Switzerland AG 2019
D. Radojčić, *Reflections on Power Prediction Modeling of Conventional High-Speed Craft*, SpringerBriefs in Applied Sciences and Technology,
https://doi.org/10.1007/978-3-319-94899-7_5

The usual design practice is to calculate the appendage resistance (shaft, strut, rudder, skeg etc.), as:

1. A percentage of bare hull resistance, or from a simple expression for all appendages (see Müller-Graf 1981; Blount and Fox 1976, respectively).
2. From expressions which calculate a component of each appendage separately (most frequently used are those published in Hoerner 1965 and combined into a complete method by Hadler 1966; see also Lasky 1980). For rudder resistance see for instance Molland and Turnock (2007) and Gregory and Dobay (1973). When the appendage configuration resembles that of an outboard or sterndrive propulsion (i.e. surface piercing drive systems); see Scherer and Patil (2011).
3. Data from model experiments (in which case the scale of the model, i.e. Reynolds number, may be important). Test results of typical full-scale appendages are given in Gregory and Beach (1979).

A procedure for predicting HSC turning characteristics is given in Denny and Hubble (1991) and Lewandowski (1993); see also Bowles (2012).

Spray Rails, Wedges, Flaps, Interceptors

Wedges, Flaps, Interceptors and Spray Rails (or spray strips) may be treated as appendages with negative resistance (i.e. if well-designed, overall resistance reduction can be achieved).

Spray Rails

Spray rails (i.e. longitudinal steps) when fitted on the bottom (near and below calm water surface), reduce the wetted area and are beneficial at higher planing speeds (see Clement 1964; Grigoropoulos and Loukakis 1995). Somewhat different spray rails (i.e. spray separators or deflectors, acting as side bow knuckles) fitted above or near the waterline reduce spray of semi-displacement hulls (particularly for $Fn_L > 0.7$) but also increase dynamic trim. The *spray rail system* for semi-displacement vessels (spray rails in combination with wedge) that reduce resistance by 6–10% in the speed range $Fn_L = 0.5$–0.9 and increase dynamic stability is proposed by Müller-Graf (1991).

Stern Wedges and Flaps

Stern wedges (buttocks' aft hook), flaps, and interceptors are beneficial in the resistance-hump region (i.e. above $Fn_L \approx 0.5$) since they increase hydrodynamic lift and reduce the resistance and dynamic trim. Flap and wedge design references are Brown (1971), Savitsky and Brown (1976) and Chen et al. (1993), and Millward (1976), for planing and round bilge hulls, respectively. *Integrated wedge-flap* combination for fast displacement and semi-displacement vessels with reported power savings of up to 12% are investigated in Cusanelli and Karafiath (1997).

Interceptors

Interceptors are nowadays used as an alternative to wedges and flaps, see for instance De Luca and Pensa (2012), where it is concluded that: (a) Interceptors are beneficial for $Fn_V = 1.8$–2.4; and (b) Savings of 15% R_T are feasible. Flaps and interceptors are movable/controllable devices and are often used as ride control appliances to improve seakeeping performance.

Dynamic Instability

Note that planing-craft-specific longitudinal instability (i.e. pitch-heave instability), often called porpoising, may be reduced or prevented by trim reduction. Caution is necessary however, as transverse dynamic stability may be degraded at low trim due to generation of a low dynamic pressure on the bow. Risk of dynamic instability for round bilge and hard chine craft should be considered for Fn_L larger than 0.75 and 0.95, respectively (Blount and McGrath 2009). Recommendations for avoidance of dynamic instability are given in Blount and Codega (1992) and Müller-Graf (1997c). A method to evaluate coupled roll-yaw-sway dynamic stability of planing craft is given in Lewandowski (1997). A useful overview of planing hull transverse dynamic stability is given in Ruscelli et al. (2012). The prediction of porpoising inception for planing craft is elaborated in Celano (1998). These methods should be integrated in the power prediction routines.

5.2.2 Air and Wind Resistance

Still-air drag (resistance) may be evaluated from:

1. A simple expression as a function of above-water projected frontal area, superstructure shape (depends on aerodynamics or superstructure streamlining), and aerodynamic drag coefficient (which has to be estimated), or
2. A percentage of total bare hull resistance [e.g. $C_{AA} = 2$–3% R_T to even 9% R_T at speeds of 30+ knots, see Müller-Graf (1981) and Blount and Bartee (1997), respectively]. Air resistance is also discussed in Blount (2014).

Wind resistance however, may be a much larger additional component, which depends on wind velocity and wind direction relative to the vessel direction (peak values are for relative headwind of around 30°). Note that other underwater components also increase due to wind-induced waves.

A noteworthy reference on this topic is Fossati et al. (2013). It suggests that the aerodynamic forces should not be accounted for just as additional resistance components, but that they should be included in the equations of equilibrium (discussed later in Sect. 6.1), since they impact both the resistance and the dynamic trim. Non-dimensional coefficients for evaluation of air and wind resistance of high speed motor yachts with standard superstructure profile are provided in the same reference. Mega yacht aerodynamics is treated in Fossati et al. (2014).

5.2.3   Correlation Allowance and Margins

Correlation allowance (C_A) should account for the differences between the ideal towing tank conditions (hence also the predictions of the MMs), and full-scale real conditions. For the slower HSC, i.e. the displacement and semi-displacement vessels, C_A depends on the extrapolation method used (ATTC-1947, ITTC-1957, ITTC-1978), and on whether the resistance components of lesser magnitude are accounted for separately (e.g. added resistance due to course keeping, spray wetted area etc.). Usually, within a correlation allowance a small allowance is accounted for, resulting in C_A being on the conservative side. For better insights see Müller-Graf (1997b). For vessels less than 80 m a correlation allowance of 0.4×10^{-3} approximately corresponds to a power increase of 10% (Blount and Bjarne 1989). Note that correlation was also discussed in Sect. 3.6.

For higher HSC speeds, actually for vessels that operate through a relatively wide speed range (semi-planing and planing), the use of excess *design margin* is recommended (see Blount 2014), since a single C_A value cannot be correct over a wide speed rage (e.g. hump and top speed). That is, the standard correlation allowance procedure, derived from the conventional model-to-ship extrapolation methods, may result in insufficient margin for the hump-speed region. The solution is to use excess design margin, which need not necessarily be uniform across the entire speed range. The design margins neccesary to enable a vessel to accelerate and overcome overloading due to increase of in-service resistance is discussed in Blount and Bartee (1997). In-service margin is analogous to the power prediction factor, or load factor, usually denoted as $(1+x)$, for conventional ships.

5.3   Resistance in a Seaway

Performance of HSC in a seaway is an extremely important topic, but is beyond the scope of this work, because it requires special attention. However, for power predictions, only added resistance in a seaway is of interest. Nevertheless, MMs for powering predictions should also account for the HSC's motion (pitch, heave, accelerations etc.) which are normally a part of the basic design criteria. The simplest approach is to restrict (reduce) speed in a rough sea, although better results—i.e. enhanced HSC performance in a seaway—would be achieved if some primary and/or secondary hull characteristics are altered.

Classical references for added resistance in a seaway are Savitsky and Brown (1976), Fridsma (1971), Hoggard (1979), Hoggard and Jones (1980), etc. It should be noted, however, that the predictions of these MMs often disagree. A very useful overview of various aspects that concern seakeeping of hard chine planing craft is given in Savitsky and Koelbel (1993) and Blount (2014). HSC seakeeping and various safety, comfort, structures, machinery, etc. considerations are discussed in Faltinsen (2005). Habitability and human factors criteria, for instance, are nowadays in the focus of HSC community; see Schleicher (2008), amongst others.

5.4 Resistance in Shallow Water

Shallow water effects are noticeable whenever $h/L_{OA} < 0.80$ or $Fn_h > 0.6$–0.7. The maximum effect is in the critical region ($0.7 < Fn_h < 1.2$), where wave-making resistance (R_W) increases dramatically. Shallow water effects subside in supercritical region at $Fn_h > 1.2$, where the resistance may be lower than in deep water. Therefore, a vessel's speed in the critical region may be substantially lower than expected, and/or the power demand is substantially higher. However, when operating in the super-critical region the reverse occurs and required power to achieve a specific speed may be less than in deep water, due to reduced resistance and somewhat enhanced propulsive efficiency. Consequently, larger and faster modern commercial vessels and mega yachts are likely to experience shallow water effects and in some cases may be sailing in, what is hydrodynamically considered to be shallow or littoral waters, throughout their life.

Note that the design logic for vessels intended for littoral and shallow water operation is fundamentally different from the usual deep-water logic. Namely, HSC which must operate in all water depths (shallow and deep) must be able to operate at speeds above $Fn_L \approx 0.7$ (see Hofman and Radojčić 1997; Radojčić and Bowles 2010). At this speed, wave wash—an important aspect for the littoral environment—is inherently significant. Moreover, wave wash decay is lower in shallow water than in deep water. So, for shallow water operation it is not only the propulsor type and size that matters, it is also desirable to have a low wave-wash hull form. This calls for careful selection of speed and waterline length, although reduction of the HSC's weight is the only measure that can effectively lower both resistance and wave-wash (see Hofman and Kozarski 2000).

Shallow water resistance can be predicted within engineering accuracy, see Radojčić and Bowles (2010). The recommended approach is based on an evaluation of the ratio of shallow- to deep-water wave resistance (R_{Wh}/R_{Wd}), where the final results rely primarily on deep water data, and the possible inaccuracies (of evaluated R_{Wh}/R_{Wd}) do not influence shallow water total resistance (R_{Th}) to a great extent. This method enables an evaluation of the resistance increment in the critical region, by utilizing deep water data only, while decrements in the supercritical regime are neglected, so that the shallow water resistance predictions are on the safe side. Note however, that shallow water propulsive efficiency requires further research, i.e. water depth effects on propulsive factors are not investigated and are not sufficiently accurate. Lyakhovitsky (2007), Hofman and Radojčić (1997) and Radojčić and Bowles (2010) provide more details on shallow water effects. See Millward and Sproston (1988), and Toro (1969) and Morabito (2013) for shallow water experiments for semi-displacement and planing hull forms respectively. Self-propulsion testing of a single HSC model is discussed in Friedhoff et al. (2007).

5.5 Propulsive Coefficients

The total propulsive efficiency (η_P) consists of propeller open water efficiency (η_O), hull efficiency (η_H), relative rotative efficiency (η_R), and shaft (including a gearbox) efficiency (η_S), i.e. $\eta_P = \eta_H \cdot \eta_R \cdot \eta_S \cdot \eta_O$; see for instance Harvald (1983), or van Manen and van Oossanen (1988). Hull efficiency can be further fragmented and expressed as $\eta_H = (1 - t)/(1 - w)$, where "t" and "w" are thrust deduction fraction and wake fraction, respectively. Similarly, $\eta_R = \eta_B/\eta_O$, where η_B is propeller efficiency behind the vessel. Consequently, for power prediction it is necessary to evaluate overall propulsive efficiency η_P, which consists of propulsive coefficients $\eta_H = f(t, w)$, $\eta_R = f(\eta_O, \eta_B)$, η_S, and η_O.

Note that the hull-propeller interaction elaborated above is correct for conventional ships (displacement vessels). Hull-propeller interaction for HSC is much more complex (hence the need for an integrated aproach) because:

1. Water inflow to the propeller is usually not axial (as is for displacement vessels), but is oblique [i.e. $v_a \cdot \cos(\psi + \tau)$], due to shaft inclination relative to hull (ψ) and dynamic trim (τ); see Fig. 6.1, and
2. Cavitation effects.

Moreover, correct "behind" condition testing under cavitation conditions is still not feasible, so that an alternative approach is required. Two separate types of experiments are usually performed—one under atmospheric (in the conventional towing tank), and one under depressurised conditions (in the cavitation tunnel/channel); see ITTC (1984). As a consequence propulsive coefficients for HSC, particularly for the cavitation regime, are not as reliable as those for conventional ships.

MMs for prediction of open water efficiency are already discussed in Sect. 4.2, while other propulsive coefficients for displacement, semi-displacement, and planing craft are given in Blount and Bjarne (1989). See Blount (1997) if the HSC's propellers are recessed into tunnels, and hence a reduction of shaft inclination is enabled. See Katayama et al. (2012) for outboard and sterndrive configurations. HSC propulsive coefficients for various propeller configurations are also given in Blount (2014). Influence of cavitation on the propeller-hull interaction is elaborated in Rutgersson (1982).

Regression analysis is employed for mathematical modeling of propulsive coefficients for high-speed round bilge hulls tested in the speed range of $Fn_L = 0.4$–1.05 and conventional installations where power is delivered to the propeller along an inclined shaft; see Bailey (1982). Independent variables are standard high-speed round bilge hull and propeller parameters. MMs are simple with 4–8 terms only. For each propulsive coefficient two equations are derived for $C_B < 0.45$ and for $C_B = 0.45$–0.51, matching faster and slower hulls respectively.

When published data is used, attention should be paid to the proper interpretation of propulsive coefficients, and particularly to the thrust deduction factor (t). Shaft inclination for instance, influences propeller characteristics more than hull-propeller interaction, and yet this influence is often expressed solely by the propulsive coeffi-

Table 5.1 Recommended references for evaluation of Additional resistance components and Propulsive coefficients

	Reference	
ADDITIONAL RESISTANCE COMPONENTS	**Calm and deep water**	
	Appendages	
	Struts, Shafts etc.	Hadler (1966), Lasky (1980)
	Rudders[a]	Gregory and Dobay (1973)
	Sterndrives, Outboards	Scherer and Patil (2011)
	Model experiments	Gregory and Beach (1979)
	Flaps and wedges[b]	Brown (1971)—planing hulls
		Millward (1976)—semi-displacement
	Interceptors[b]	De Luca and Pensa (2012)
	Air resistance	
	Blount (2014), Fossati et al. (2013)	
	Correlation allowance and margins	
	Blount (2014)	
	Resistance in a seaway	
	Savitsky and Brown (1976)	
	Hoggard (1979)	
	Hoggard and Jones (1980)	
	Blount (2014)	
	Resistance in shallow water	
	Radojčić and Bowles (2010)—MM	
	Millward and Sproston (1988)—tests semi-displacement	
	Toro (1969)—tests planing hulls	
	Friedhoff et al. (2007)—self-propulsion tests, planing	
PROPULSIVE COEFFICIENTS	Blount and Bjarne (1989)—HSC, all kind	
	Bailey (1982)—Round bilge	
	Blount (1997)—Propeller in tunnel	
	Katayama et al. (2012)—Outboards, Sterndrives	

[a]Maneuverability and Turning characteristics:
– Denny and Hubble (1991), Lewandowski (1993)

[b]Reduce dynamic trim and may cause dynamic instability:
– Transverse (dynamic trim too low)—Blount and Codega (1992), Müller-Graf (1997c)

– Longitudinal, porposing (dynamic trim too high)—Celano (1998)

cients. Therefore, the specific conditions upon which the propulsive coefficients are based should be clarified (e.g. whether the t value is valid for the horizontal resistance component or inclined ($\psi + \tau$) thrust vector, or whether the appendage resistance is accounted for, etc.); see van Manen and van Oossanen (1988) and Blount (2014).

5.6 Recommended References for Evaluation of Additional Resistance Components and Propulsive Coefficients

This section provides an overview of references linked to the:

• Additional resistance components (related to bare hull resistance in calm and deep water), and
• Propulsive coefficients (other than open water efficiency).

Suggested references are presented in Table 5.1.

References

Bailey D (1982) A statistical analysis of propulsion data obtained from models of high speed round bilge hulls. In: RINA Symposium on Small Fast Warships and Security Vessels, London

Blount DL (1997) Design of propeller tunnels for high speed craft. In: Proceedings of the 4th International Conference on Fast Sea Transportation (FAST '97), Sydney

Blount DL (2014) Performance by design. ISBN 0-978-9890837-1-3

Blount DL, Bartee RJ (1997) Design of propulsion systems for high-speed craft. Mar Technol 34(4)

Blount DL, Bjarne E (1989) Design and selection of propulsors for high speed craft. In: 7th Lips Propeller Symposium, Nordwijk-on-Sea

Blount DL, Codega LT (1992) Dynamic stability of planing boats. Mar Technol 29(1)

Blount DL, Fox DL (1976) Small craft power prediction. Mar Technol 13(1)

Blount DL, McGrath JA (2009) Resistance characteristics of semi-displacement mega yacht hull forms. RINA Trans, Int J Small Craft Technol 151(Part B2)

Bowles J (2012) Turning characteristics and capabilities of high speed monohulls. In: SNAME's 3rd Chesapeake Power Boat Symposium, Annapolis

Brown PW (1971) An experimental and theoretical study of planing surfaces with trim flaps. Davidson Laboratory Report 1463, Stevens Institute of Technology

Celano T (1998) The prediction of porpoising inception for modern planing craft. USNA Trident Report No. 254. Annapolis

Chen CS, Hsueh TJ, Fwu J (1993) The systematic test of wedge on flat plate planing surface. In: Proceedings of the 2nd International Conference on Fast Sea Transportation (FAST '93), Yokohama

Clement EP (1964) Reduction of planing boat resistance by deflection of the whisker spray. DTMB Report 1920

Cusanelli D, Karafiath G (1997) Integrated wedge-flap for enhanced powering performance. In: Proceedings of the 4th International Conference on Fast Sea Transportation (FAST '97), Sydney

De Luca F, Pensa C (2012) Experimental investigation on conventional and unconventional interceptors. RINA Trans, Int J Small Craft Technol 153(Part B2)

Denny SB, Hubble EN (1991) Predicting of craft turning characteristics. Mar Technol 28(1)

Faltinsen OM (2005) Hydrodynamics of high-speed marine vehicles. Cambridge University Press. ISBN-13 978-0-521-84568-7

Fossati F, Muggiasca S, Bertorello C (2013) Aerodynamics of high speed small craft. In: Proceedings of the 12th International Conference on Fast Sea Transportation (FAST 2013), Amsterdam

Fossati F, Robustelli F, Belloli M, Bertorello C, Dellepiane S (2014) Experimental assessment of mega-yacht aerodynamic performance and characteristics. RINA Trans 156(Part B2). London. https://doi.org/10.3940/rina.ijsct.2014.b2.157

Fridsma G (1971) A systematic study of the rough water performance of planing boats (Irregular waves—Part II). Davidson Laboratory Report 1495

Friedhoff B, Henn R, Jiang T, Stuntz N (2007) Investigation of planing craft in shallow water. In: Proceedings of the 9th International Conference on Fast Sea Transportation (FAST 2007), Shanghai

Gregory D, Beach T (1979) Resistance measurements of typical planing boat appendages. DTNSRDC Report SPD-0911-01

Gregory DL, Dobay GF (1973) The performance of high-speed rudders in a cavitating environment. SNAME Spring Meeting, Florida

Grigoropoulos GJ, Loukakis TA (1995) Effect of spray rails on the resistance of planing hulls. In: Proceedings of the 3rd International Conference on Fast Sea Transportation (FAST '95), Lubeck-Travemunde

Hadler JB (1966) The prediction of power performance of planing craft. SNAME Trans 74

Harvald SA (1983) Resistance and propulsion of ships. Ocean Engineering (Wiley). ISBN 0471063533

Hoerner SF (1965) Fluid dynamic drag. Book published by the author, Midland Park

Hofman M, Kozarski V (2000) Shallow water resistance charts for preliminary vessel design. Int Shipbuild Prog 47(449)

Hofman M, Radojčić D (1997) Resistance and propulsion of fast ships in shallow water. Faculty of Mechanical Engineering, University of Belgrade, Belgrade. ISBN 86-7083-297-6 (in Serbian)

Hoggard MM (1979) Examining added drag of planing craft operating in the seaway. Hampton Road Section of SNAME

Hoggard MM, Jones MP (1980) Examining pitch, heave and accelerations of planing craft operating in a seaway. In: High Speed Surface Craft Conference, Brighton

ITTC (1984) Proceedings of the 17th International Towing Tank Conference, High-Speed Propulsion, vol 1, Goteborg

ITTC (2011) Recommended procedures and guidelines—High speed marine vehicles (Section 7.5-02-05)—Resistance test, Section 7.5-02-05-01; Propulsion test Section 7.5-02-05-02

Katayama T, Nishihara Y, Sato T (2012) A study on the characteristics of self-propulsion factors of planing craft with outboard engine. In: SNAME's 3rd Chesapeake Power Boat Symposium, Annapolis

Lasky MP (1980) An investigation of appendage drag. DTNSRDC Report SPD-458-01

Lewandowski E (1993) Manueverability of high-speed power boats. In: 5th Power Boat Symposium, SNAME Southeast Section

Lewandowski E (1997) Transverse dynamic stability of planing craft. Mar Technol 34(2)

Lyakhovitsky A (2007) Shallow water and supercritical ships. Backbone Publishing Company, Hoboken

Millward A (1976) Effect of wedges on the performance characteristics of two planing hulls. J Ship Res 20(4)

Millward A, Sproston J (1988) The prediction of resistance of fast displacement hulls in shallow water. RINA Maritime Technology Monograph No. 9, London

Molland AF, Turnock SR (2007) Marine rudders and control surfaces—principles, data, design and applications. Elsevier. ISBN 978-0-75-066944-3

Morabito MG (2013) Planing in shallow water at critical speed. J Ship Res 57(2)

Müller-Graf B (1981) Semidisplacement round bilge vessels. In: Status of hydrodynamic technology as related to model tests of high speed marine vehicles (Section 3.2), DTNSRDC Report 81/026

Müller-Graf B (1991) The effect of an advanced spray rail system on resistance and development of spray on semi-displacement round bilge hulls. In: Proceedings of the 1st International Conference on Fast Sea Transportation (FAST '91), Trondheim

Müller-Graf B (1997a) Part I: Resistance components of high speed small craft. In: 25th WEGEMT School, Small Craft Technology, NTUA, Athens. ISBN I 900 453 053

Müller-Graf B (1997b) Part III: Factors affecting the reliability and accuracy of the resistance prediction. In: 25th WEGEMT School, Small Craft Technology, NTUA, Athens. ISBN I 900 453 053

Müller-Graf B (1997c) Dynamic stability of high speed small craft. In: 25th WEGEMT School, Small Craft Technology, NTUA, Athens. ISBN I 900 453 053

Radojčić D, Bowles J (2010) On high speed monohulls in shallow water. In: SNAME's 2nd Chesapeake Power Boat Symposium, Annapolis

Ruscelli D, Gualeni P, Viviani M (2012) An overview of planing monohulls transverse dynamic stability and possible implications with static intact stability rules. RINA Trans 154(Part B2). London. https://doi.org/10.3940/rina.ijsct.2012.b2.134

Rutgersson O (1982) High speed propeller performance—influence of cavitation on the propeller-hull interaction. Ph.D. thesis, Chalmers University of Technology, Goteborg. ISBN 91-7032-072-1

Savitsky D (1981) Planing hulls. In: Status of hydrodynamic technology as related to model tests of high speed marine vehicles (Section 3.3), DTNSRDC Report 81/026

Savitsky D, Brown PW (1976) Procedure for hydrodynamic evaluation of planing hulls in smooth and rough water. Mar Technol 13(4)

Savitsky D, Koelbel JG (1993) Seakeeping of hard chine planing hulls. SNAME's Technical and Reasearch Panel SC-1 (Power Craft), Report R-42

Scherer JO, Patil SKR (2011) Hydrodynamics of surface piercing outboard and sterndrive propulsion systems. In: Proceedings of the 11th International Conference on Fast Sea Transportation (FAST 2011), Honolulu

Schleicher DM (2008) Regarding small craft seakeeping. In: SNAME's 1st Chesapeake Power Boat Symposium, Annapolis

Toro A (1969) Shallow-water performance of a planing boat. Department of Naval Architecture and Marine Engineering Report No. 019, Michigan University (AD-A016 682)

van Manen JD, van Oossanen P (1988) Propulsion. In: Lewis EV (ed) Principles of naval architecture, vol II, Chapter 6. SNAME, Jersey City

Chapter 6
Power Prediction

6.1 Power and Performance Predictions for High-Speed Craft

The performance prediction models are typically composed of modules (subroutines) for the calculation of bare hull resistance, appendage resistance, propeller characteristics, etc., which are derived individually and may be used independently of each other. For conventional ships the Holtrop and Mennen method (Holtrop and Mennen 1982) applicable to a wide variety of ship types is still in use despite its age, and is typically included in multiple software packages. Adequate MMs for HSC do not really exist, probably due to the unique nature of HSC, operating in displacement, semi-displacement, and often planing regimes, as discussed in Sect. 1.2. Specifically, with increasing speed HSC change both the displacement and trim, which is not the case with conventional (displacement) ships. Therefore, in order to model the HSC's operating conditions and power requirement, equations of equilibrium must be formed.

Equations of Equilibrium

Note that the equations of equilibrium (including interpretation of propeller forces), are closely connected with propulsive coefficients, and in particular with propulsive efficiency η_P and hence also power prediction. That is, in general

$$T_h = T_a \cdot \cos(\psi + \tau) - N \cdot \sin(\psi + \tau) \quad \text{and} \quad L = T_a \cdot \sin(\psi + \tau) + N \cdot \cos(\psi + \tau),$$

where T_h, L, T_a, and N are the propeller's horizontal force, lift (vertical force), axial force, and normal force, respectively; see Fig. 6.1. Basically, three different approaches of accounting for propeller forces may be considered:

1. Shaft inclination ($\psi + \tau$) is neglected. Simplification of this kind is permissible for $\psi < 6°-8°$ (Blount and Bjarne 1989), i.e. for the displacement vessels when $\psi + \tau$ is small. Nevertheless, this is often, albeit wrongly, done even when $\psi > 6°-8°$. This simplification also misinterprets propeller-hull interaction with altered propeller characteristics due to oblique inflow.
2. Only the propeller's horizontal forces are accounted for, through $T_h = T_a \cdot \cos(\psi + \tau) - N \cdot \sin(\psi + \tau)$, while vertical force L (lift) is neglected. This is a more realistic approach than the previous one, and most power prediction routines use it.
3. Both the propeller's horizontal T_h and vertical force L are accounted for through the equilibrium equations. This means that the propeller's vertical force L reduces both the displacement and the trim, which further impacts the resistance and significantly complicates power predictions. In addition, the ratio of T_a/N for non-cavitating and cavitating conditions is not the same, i.e. N decreases as cavitation increases (see ITTC 1984), so $(T_a/N)_{CAV} > (T_a/N)_{NON-CAV}$, which additionally complicates the analysis.

This discussion resembles the one presented in ITTC (1984), where model testing, high-speed propulsion, inclined shaft propeller forces, cavitation effects on propulsive efficiency etc., were elaborated. Accordingly, model experiments where all influences are taken into account (corresponding to Approach 3) are difficult to execute; see discussion in Sect. 5.5. Simpler approaches for model testing, and consequently for mathematical modeling, corresponding to methods 1 and 2 are usually practiced, and are often combined with some correlation factors.

Finally, it may be concluded that the interaction between the resistance and the propulsion modules, expressed through the equations of equilibrium and propulsive coefficients, in essence differentiates the powering prediction concepts among themselves.

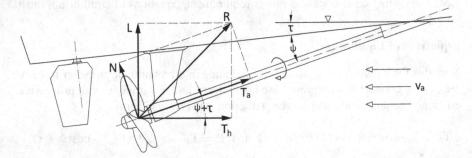

Fig. 6.1 Propeller forces on inclined shaft

6.2 Classics[1]

Papers in this category are written before the wide use of computers in everyday engineering practice. They include graphs and tables, and often instructions/routines for hand-held calculators. Nevertheless, in the author's opinion these papers have "enduring quality" (i.e. value) and contain information necessary for understanding HSC hydrodynamics. They are therefore highly recommended as background reading, even for those who rely on commercial software in their design work.

Hadler (1966)

Equations of equilibrium are examined in Hadler (1966). Propeller forces on an inclined shaft were superimposed on the bare hull planing hydrodynamics (in essence following Savitsky's method for resistance evaluation). Appendage lift and drag were examined too (skeg, rudders, shafts, struts etc.), since it was necessary to balance all relevant forces in the equilibrium equations. The methodology for performance analysis was established by examining the interaction of propeller and planing surface forces. Amongst the conclusions was that in "optimizing the design of high-performance planing hull, the whole hydrodynamic system must be considered".

Hadler and Hubble (1971)

Hadler's approach was soon upgraded (Hadler and Hubble 1971) and prismatic hulls and Troost open water propeller series were replaced with the Series 62 (12.5° deadrise) and Gawn-Burrill KCA propeller series respectively. Single, twin, and quadruple-screw configurations were investigated for wide speed and hull size ranges, giving optimum diameters versus RPMs, optimum propulsive coefficients etc. The charts presented are especially useful to planing hull designers for power prediction decisions typical in the preliminary design phases.

Blount and Fox (1976)

The technical data available at the time has been further organized in Blount and Fox (1976), resulting in a methodology for hard chine planing craft power prediction. This power prediction method accounts for bare hull resistance, various appendage configurations, propeller characteristics under cavitating and non-cavitating conditions, and resistance augmentation due to rough water. For bare hull resistance evaluation, Savitsky's method for prismatic hulls was modified with the so called *multiplying factor M*. The *effective beam and deadrise* were also investigated for non-prismatic hulls. Propeller selection procedure, beside the Gawn-Burrill series, allowed the use of other propeller series (Newton-Rader for instance). For rough water performance the Fridsma (1971) method was applied.

---

[1]The definition for the *classic literature*, according to Dictionary.com, is: "An author or literary work of the first rank, especially one of demonstrably enduring quality". Here, word *classics* should be replaced with *classic papers, references* or *classic power prediction approaches.*

Savitsky and Brown (1976)[2]

Studies of planing hull hydrodynamics, conducted by the recognized Davidson Laboratory, are presented in Savitsky and Brown (1976). MMs for resistance prediction for the pre-planing and full planing regimes were pulled from Mercier and Savitsky (1973) and Savitsky (1964) respectively, with the addition of bottom warp discussion. An in-depth study of the effects of bottom warp are given four decades later in Savitsky (2012) and in Begović and Bertorello (2012). Seakeeping prediction is according to the Fridsma (1971) method, and the effectiveness of trim control flaps according to Brown (1971).

Müller-Graf (1997a, b, c, d)

In spite of being presented well within the "computer era", Müller-Graf's WEGEMT Lectures on Small Craft Technology (Müller-Graf 1997a, b, c, d) are certainly amongst the unavoidable classics. Namely, in Part I are various HSC resistance components and subcomponents under trial conditions and procedures for their determination. Part II treats published systematical resistance data and deals with power predictions of different HSC types. Part III discusses various uncertainties of prediction methods and suggests appropriate safety margins. Different types of dynamic instabilities and measures to improve them are discussed in Müller-Graf (1997d).

6.3 Modernism[3]

MMs presented in this section are developed essentially with computer application in mind. Moreover, the aim is often not only to predict the performance of a HSC with given principal characteristics, but also to enable the designer to choose the optimal characteristics of the hull and of the propulsor during the preliminary design stages. In general this might be done through the:

1. Systematic variation of input variables (i.e. through a parametric study, nowadays using a simple spread-sheet program)—for which the user's interaction is necessary to some extent.
2. Application of general nonlinear constrained optimization techniques. This, however, requires strict definition of MM's applicability bounds, and hence these bounds practically became a part of the MM itself. Consequently, the complete MM consists of an objective function for performance prediction and the constraints—which are usually nonlinear equations of equality and inequality type.

---

[2]Note that Savitsky and Brown (1976) treats hull hydrodynamics only, and is an extension to prismatic hull method approach. That is, propulsion is not discussed, and hence, strictly speaking, this work doesn't belong in the "Power prediction" category.

[3]The definition for *modernism*, according to Merriam-Webster.com, is: "A style of art, architecture, literature, etc., that uses ideas and methods which are very different from those used in the past". Here, the word *modernism* should be replaced with *modern power prediction approaches*.

Note that the comprehensive MM often must be presented in a form which can be efficiently handled by the chosen optimization technique (see Radojčić 1985a for instance, where the optimization method used was *Sequential Unconstrained Minimization Technique—SUMT*).

Four MMs, or approaches, with the aim to minimize power, rather than to minimize the resistance and maximize efficiency of propellers, are mentioned here.

Hubble (1980)

This paper gives a procedure for prediction of power and vertical accelerations for planing hulls, with propellers on inclined shafts. Speed versus wave heights is developed based on the power limits of the propulsion system, and endurance limits of the crew due to vertical acceleration. Power optimization is done through a systematic variation of input variables, i.e. it is a parametric study. To evaluate power, several independent state-of-the-art modules were used: for prediction of bare hull resistance (Hubble 1974); appendage resistance and propulsive coefficients (Blount and Fox 1976); propeller characteristics (Oosterveld and van Oossanen 1975; Blount and Fox 1978); habitability limits (Hoggard and Jones 1980) etc. A much broader planing craft feasibility model, which in addition to power prediction, incorporates weight groups, structures, engines, loads etc., is explained in Hubble (1978). Both procedures, however, are based on an approach that is nowadays somewhat obsolete, since computer power is used mainly for the interpolations etc. This is rather similar to what would have been done "by hand" in the pre-computer era.

Calkins (1983)

This routine is developed for recreational powerboats and is based on the design spiral concept, where powering "absorbs" just a few modules (a phrase *technological areas* was used). That is, the abovementioned Hubble (1978, 1980) MMs consisted of 7 and 10 modules respectively, while this procedure consists of 10 modules. However, since the module's scope is not identical, direct comparison does not make much sense. This is actually an interactive CAD synthesis program, which inherently relies much more on computers and graphics than previous instances. In summary, although powerboat modules and sub-modules are simple equations, the logic of this program essentially corresponds to present-day CAD methods.

Radojčić (1991)

The primary objective of this paper is to present a new methodology for planing craft power prediction. A nonlinear constrained optimization technique is used, and design optimization is done on the whole system (min P_D), rather than on any of the components (min R_T and max η_D). Power prediction modules (subroutines) are those for: bare hull resistance (Radojčić 1985b), appendage resistance (Savitsky and Brown 1976), rough water performance (Hoggard 1979; Hoggard and Jones 1980), KCA propeller performance for open-water and cavitating environment (Radojčić 1988b), and propeller performance on inclined shaft (Radojčić 1988a, based on Gutche's quasi-steady method[4]—Gutsche 1964 and Taniguchi et al. 1967 for respectively, non-cavitating and cavitating propellers). The equilibrium equations (essentially similar to those in Hadler 1966) articulate the interaction between the planing craft's resistance and propulsion forces. This, to a certain extent reduces the importance of propulsive coefficients as Approach 3 explained in Sect. 6.1 is used (i.e. both propeller's horizontal T_h and vertical force L are taken into account through the equilibrium equations). Since several hull and propeller variables have to be examined simultaneously, it is necessary to apply one of the optimization routines. For this purpose the least efficient method is chosen: the Monte Carlo technique (particularly, *multistage Monte Carlo integer optimization technique*; see Conley 1981), mainly because it is simple, robust and easy to program. The goal is to choose the best nonfixed parameters of the hull and propulsors from the performance point of view (min P_D) and then to conduct a tradeoff study, which is important in the early design process.

Use of general optimization routines, as explained above, is not so common; improvements may be implemented in the following three areas:

1. Existing modules (individual MMs) within the synthesis routine can be replaced with better ones, or new modules could be added. This would also enhance the power prediction routine. Modules are important because they shape the prediction. Individual modules are in fact the focus of present work and have therefore been discussed in the previous sections.
2. The interaction between the modules, expressed by the equations of equilibrium and hull-propeller interaction in non-cavitating and cavitating conditions, are the next area of possible enhancements. Also, introduction of correlation factors seems to be inevitable.
3. The optimization technique itself is a third important part of the powering prediction routines, because it provides the tradeoff information and enables the choice of optimal characteristics of the hull and of the propulsor. A short and informative article about optimization is given in Collete (2014).

Although discussion about optimization techniques is beyond the scope of present work, two recent references will be mentioned:

– Mohamad Ayob et al. (2011) showing the application of a single-objective minimization technique for calm-water resistance and multiobjective minimization of total resistance, steady turning diameter and vertical impact acceleration.
– Knight et al. (2014) using a multiobjective optimization technique to minimize resistance and vertical acceleration.

Incidentally, in both cases the Savitsky (1964) method is used for calm-water planing hull resistance evaluation.

---

[4]This method allows recalculation of the commonly available data for axial water inflow to those that correspond to oblique inflow conditions (i.e. propeller on inclined shaft). In essence, water inflow velocity is assumed to be $v_{OBLOUE} = v_{AXIAL} \cdot \cos(\psi + \tau)$ and the propeller's inclined shaft unsteady (oscillatory) forces are averaged across one revolution.

Müller-Graf et al. (2003a, b)

Series 89' is different from everything discussed up to here, not only because it deals with high-speed, hard chine, planing catamaran hulls, but also because extensive resistance and self-propulsion tests were carried out (see Table 3.1). An enormous quantity of data was gathered through resistance and self-propulsion tests carried out with relatively large models. An immense stock of data is represented by regression derived MMs for resistance, trim, propulsion coefficients, and delivered power. The tests and results were validated for trial conditions and thus are believed to be very reliable. Prior to this modeling, Zips (1995) presented a regression analysis for evaluation of the Series 89' residuary resistance. Zips' mathematical model is simpler and also less accurate than the MMs presented in Müller-Graf et al. (2003a).

Through the application of regression analysis, three groups of mathematical models for the reliable evaluation of power have been derived:

- Speed-independent model for the residuary resistance-to-weight ratio $\varepsilon_R = R_R/\Delta \cdot g = f(\beta_M, \delta_W, L_{WL}/B_{XDH})$,
- Deadrise-independent models for the evaluation of dynamic trim (denoted here θ), propulsive coefficients, and specific power ratio (i.e. θ, η_D, η_O, η_H, w, t and $\varepsilon_B = P_B/\Delta \cdot g \cdot v$), all as $f(Fn_{\nabla/2}, L_{WL}/B_{XDH})$,
- Models for wetted surface coefficient (WSC $= S/(\nabla/2)^{2/3}$), and length-to-displacement ratio $(L_{WL}/(\nabla/2)^{1/3})$, as a $f(\beta_M, L_{WL}/B_{XDH})$,

where β_M—angle of deadrise amidship, δ_W—angle of transom wedge, L_{WL}/B_{XDH}—ratio of waterline length and maximum beam of demihull.

The power prediction method is reliable because the propulsive coefficients for Series 89' were obtained from the self-propulsion tests under trial conditions (i.e. the model propellers are working at a loading which is equal to that of the full-scale propellers). Therefore, none of the less accurate speed-independent trial allowances for power and RPM have to be applied to the power obtained by these regression models. The method also enables the optimization of hull-form parameters and prediction of power in early design stage for catamarans with length-to-beam ratios of $L_{WL}/B_{XDH} = 7.55-13.55$ and midship deadrise values of $\beta_M = 16°-38°$ with an optimal wedge inclination of $\delta_W = 8°$ Residuary resistance and hull-propeller interaction coefficients (i.e. propulsive coefficients) versus speed are given for a large speed range from the hump speed to planing speeds (up to $Fn_{\nabla/2} = 4$ or $Fn_L \approx 1.5$). Power may be evaluated by: (a) conventional method ($P_{BTR} = R_{TTR} \cdot v /(\eta_D \cdot \eta_S)$), or (b) short method ($P_{BTR} = \varepsilon_{BTR} \cdot \Delta \cdot g \cdot v$).

Note that the regression models for propulsive coefficients may be used relatively successfully for other catamaran forms, as propulsive coefficients are influenced more by length-to-beam ratio than by section shape. So, power predictions for round bilge hull form catamarans are possible if accurate resistance characteristics are available. The Southampton round bilge catamaran series[5] is by far the best known, see Molland et al. (1996) and Molland and Lee (1997), for instance, or Molland et al. (2011); resistance prediction for this series was also mentioned in the Sects. 2.2.2 and 3.4.12.

6.4 Another Perspective[6]

As stated in the Introduction, *MM development is an evolutionary process.* Namely, new MMs are expected to be better than the previous ones. In general this pattern is followed and the results of any new procedures are compared with the previous ones. For example, the author used an ANN technique first in Radojčić et al. (2014a, b), but cautiously and together with already proven regression analysis. Since encouraging results were obtained, modeling was executed solely with ANN in Radojčić et al. (2017). In that paper single- and multiple-output ANN routines were compared. Subsequently, only a multiple output ANN was used in Radojčić and Kalajdžić (2018) for derivation of MMs for R/Δ & τ and $S/\nabla^{2/3}$ & L_{WL}/L_P. This historical perspective is illustrated in Fig. 6.2, showing a timeline of some notable MMs along with their main attributes. In the author's opinion the MMs shown are important milestones that have left a footprint on MM development methodologies.

---

[5]Couser et al. (1997) paper entitled "Calm water powering predictions for high speed catamarans", actually treats various resistance components, extrapolation methods for catamarans etc., not powering as designated here.

[6]The definition for the *perspective*, according to Cambridge English Dictionary is: "A particular way of viewing things that depends on one's experience and personality". Here, the word *perspective* should be replaced with *additional comment about mathematical modeling*.

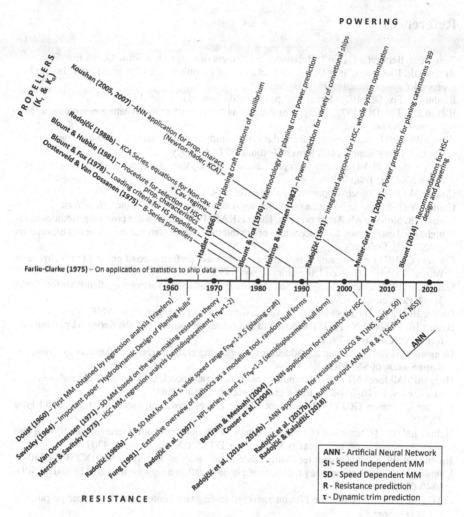

POWERING

PROPELLERS (K_T & K_Q)

Koushan (2005, 2007) – ANN application for prop. charact.

Radojčić (1988b) – KCA Series, equations for Non-cav. + Cav. regime

Blount & Hubble (1981) – KCA Series, equations for Non-cav.

Blount & Fox (1978) – Procedure for selection of HSC prop. characteristics

Oosterveld & Van Oossanen (1975) – B-Series propellers

Hadler (1966) – Procedure for selection of HSC prop. characteristics

First planing craft equations of equilibrium (Newton-Rader, KCA)

Blount & Fox (1976) – Methodology for planing craft power prediction

Holtrop & Mennen (1982) – Power prediction for variety of conventional ships

Radojčić (1991) – Integrated approach for HSC, whole system optimization

Müller-Graf et al. (2003) – Power prediction for planing catamarans S'89

Blount (2014) – Recommendations for HSC design and powering

Farlie-Clarke (1975) – On application of statistics to ship data

1960 1970 1980 1990 2000 2010 2020

Doust (1960) – First MM obtained by regression analysis (trawlers)

Savitsky (1964) – Important paper "Hydrodynamic Design of Planing Hulls"

Van Oortmerssen (1971) – SD MM based on the wave-making resistance theory

Mercier & Savitsky (1973) – HSC MM, regression analysis (semidisplacement, $F_{n\nabla}$=1-2)

Radojčić (1985b) – SI & SD MM for R and τ, wide speed range $F_{n\nabla}$=1-3.5 (planing craft)

Fung (1991) – Extensive overview of statistics as a modeling tool, random hull forms

Radojčić et al. (1997) – NPL series, R and τ, $F_{n\nabla}$=1-3 (semidisplacement hull form)

Bertram & Mesbahi (2004) – ANN application for resistance

Couser et al. (2004) – ANN application for resistance for HSC

Radojčić et al. (2014a, 2014b) – Multiple output ANN for R & τ (Series 62, NSS)

Radojčić et al. (2017b) – ANN application for resistance (USCG & TUNS, Series 50)

Radojčić & Kalajdžić (2018)

ANN

RESISTANCE

ANN	- Artificial Neural Network
SI	- Speed Independent MM
SD	- Speed Dependent MM
R	- Resistance prediction
τ	- Dynamic trim prediction

Fig. 6.2 Timeline of some notable MMs with their principal characteristics

References

Begović E, Bertorello C (2012) Resistance assessment of warped hull form. Ocean Eng 56

Blount DL, Bjarne E (1989) Design and selection of propulsors for high speed craft. In: 7th lips propeller symposium, Nordwijk-on-Sea

Blount DL, Fox DL (1976) Small craft power prediction. Mar Technol 13(1)

Blount DL, Fox DL (1978) Design considerations for propellers in cavitating environment. Mar Technol 15(2)

Brown PW (1971) An experimental and theoretical study of planing surfaces with trim flaps. Davidson Laboratory Report 1463, Stevens Institute of Technology

Calkins DE (1983) An interactive computer aided design Synthesis program for recreational powerboats. SNAME Trans 91

Collete M (2014) Effective optimization. Mar Technol, July

Conley W (1981) Optimization, a simplified approach. Petrocelli Books Inc., New York

Couser PR, Molland AF, Armstrong NA, Utama IKAP (1997) Calm water powering predictions for high speed catamarans. In: Proceedings of 4th international conference on fast sea transportation (FAST '97), Sydney

Fridsma G (1971) A systematic study of the rough water performance of planing boats (Irregular Waves—Part II). Davidson Laboratory Report 1495

Gutsche F (1964) Untersuchung von Schiffsschrauben in schrager Austromung. Schiffbauforschung 3, 3/4, Rostock

Hadler JB (1966) The prediction of power performance of planing craft. SNAME Trans 74

Hadler JB, Hubble EN (1971) Prediction of the power performance of the Series 62 planing hull forms. SNAME Trans 79

Hoggard MM (1979) Examining added drag of planing craft operating in the seaway. Hampton Road Section of SNAME

Hoggard MM, Jones MP (1980) Examining pitch, heave and accelerations of planing craft operating in a seaway. In: High speed surface craft Conference, Brighton

Holtrop J, Mennen GGJ (1982) An approximate power prediction method. Int Shipbuild Prog 29(335)

Hubble EN (1974) Resistance of hard-chine stepless planing craft with systematic variation of hull form, longitudinal centre of gravity and loading. DTNSRDC R&D Report 4307

Hubble EN (1978) Planing craft feasibility model, user's manual. Report DTNSRDC/SPD-0840-01

Hubble EN (1980) Performance prediction of planing craft in a seaway. Report DTNSRDC/SPD-0840-02

ITTC (1984) Proceedings of the 17th international towing tank conference, High-speed propulsion, vol 1, Goteborg

Knight JT, Zahradka FT, Singer DJ, Collette MD (2014) Multiobjective particle swarm optimization of a planing craft with uncertainty. J Ship Prod Des 30(4)

Mercier JA, Savitsky D (1973) Resistance of transom-stern craft in the pre-planing regime. Davidson Laboratory Report 1667

Mohamad Ayob AF, Ray T, Smith WF (2011) Beyond hydrodynamic design optimization of planing craft. J Ship Prod Des 27(1)

Molland AF, Lee AR (1997) An investigation into the effect of prismatic coefficient on catamaran resistance. RINA Trans 139

Molland AF, Wellicome JF, Couser PR (1996) Resistance experiments on a systematic series of high speed displacement catamaran forms: variation of length-displacement ratio and breadth-draught ratio. RINA Trans 138

Molland AF, Turnock SR, Hudson DA (2011) Ship resistance and propulsion—practical estimation of ship propulsive power. Cambridge University Press, ISBN 978-0-521-76052-2

Müller-Graf B (1997a) Part I: Resistance components of high speed Small craft. In: 25th WEGEMT School, Small Craft Technology, NTUA, Athens. ISBN I 900 453 053

Müller-Graf B (1997b) Part II: Powering performance prediction of high speed small craft. In: 25th WEGEMT School, Small Craft Technology, NTUA, Athens. ISBN I 900 453 053

Müller-Graf B (1997c) Part III: actors affecting the reliability and accuracy of the resistance prediction. In: 25th WEGEMT School, Small Craft Technology, NTUA, Athens. ISBN I 900 453 053

Müller-Graf B (1997d) Dynamic stability of high speed small craft. In: 25th WEGEMT School, Small Craft Technology, NTUA, Athens. ISBN I 900 453 053

Müller-Graf B, Radojčić D, Simic A (2003a) Resistance and propulsion characteristics of the VWS hard chine catamaran hull Series 89'. SNAME Trans 110

Müller-Graf B, Radojčić D, Simic A (2003b) Discussion of paper 1: resistance and propulsion characteristics of the VWS hard chine catamaran hull Series 89'. Mar Technol 40(4)

Oosterveld MWC, van Oossanen P (1975) Further computer-analyzed data of the Wageningen B-screw series. Int Shipbuild Prog 22(251)

Radojčić D (1985a) Optimal preliminary propeller design using nonlinear constrained mathematical programming technique. University of Southampton, Ship Science Report No. 21

Radojčić D (1985b) An approximate method for calculation of resistance and trim of the planing hulls. University of Southampton, Ship Science Report No. 23. Paper presented on SNAME symposium on powerboats, Sept 1985

Radojčić D (1988a) Evaluation of propeller performance in oblique flow. In: 8th symposium on theory and practice of shipbuilding, In Memoriam of Prof. Sorta, Zagreb (in Serbian)

Radojčić D (1988b) Mathematical model of segmental section propeller series for open-water and cavitating conditions applicable in CAD. Propellerss' 88 symposium, SNAME, Virginia Beach

Radojčić D (1991) An engineering approach to predicting the hydrodynamic performance of planing craft using computer techniques. RINA Trans 133

Radojčić D, Kalajdžić M (2018) Resistance and trim modeling of Naples hard chine systematic series. RINA Trans Int J Small Craft Technol. https://doi.org/10.3940/rina.ijsct.2018.b1.211

Radojčić D, Zgradić A, Kalajdžić M, Simić A (2014a) Resistance prediction for hard chine hulls in the pre-planing regime. Pol Marit Res 21(2(82)), Gdansk

Radojčić D, Morabito M, Simić A, Zgradić A (2014b) Modeling with regression analysis and artificial neural networks the resistance and trim of Series 50 experiments with V-bottom motor boats. J Ship Prod Des 30(4)

Radojčić DV, Kalajdžić MD, Zgradić AB, Simić AP (2017) Resistance and trim modeling of systematic planing hull Series 62 (With 12.5, 25 and 30 degrees deadrise angles) using artificial neural networks, Part 2: Mathematical models. J Ship Prod Des 33(4)

Savitsky D (1964) Hydrodynamic design of planing hulls. Mar Technol 1(1)

Savitsky D (2012) The effect of bottom warp on the performance of planing hulls. SNAME's 3rd Chesapeake Power Boat Symposium, Annapolis

Savitsky D, Brown PW (1976) Procedure for hydrodynamic evaluation of planing hulls in smooth and rough water. Mar Techno 13(4)

Taniguchi K, Tanibayashi H, Chiba N (1967) Investigation into the propeller cavitation in oblique flow. Mitsubishi Technical Bulletin No. 143

Zips JM (1995) Numerical resistance prediction based on the results of the VWS hard chine catamaran hull Series 89'. In: Proceedings of 3rd international conference on fast sea transportation (FAST '95), Lübeck-Travemünde

Chapter 7
Concluding Remarks

This work is an overview that outlines the author's views. It is believed to be of value for MM developers and the HSC community. In addition, key references important for HSC hydrodynamics, and in particular for resistance, propulsion, and power prediction, are provided.

MMs for resistance prediction are the core of present work. The MMs presented here address conventional HSC, which may be classified as those belonging to the semi-displacement type (NPL, VTT, SKLAD, NTUA) and planing type (Series 50, 62, 65-B, TUNS, USCG, NSS). For the first group, the length Froude number (Fn_L) up to 1.1 is of interest. For the second group the volumetric Froude number (Fn_∇) up to 5.5 or so is appropriate. Adequate resistance and trim MMs for small HSC (boats) for speeds corresponding to Fn_∇ of up to 8—for which the stepped hull form would be advantageous—is missing.

In order to achieve better accuracy, contemporary MMs are complex, and include approximately 10 times as many equation terms compared to those created a few decades ago. Note however, that the number of base parameters (input variables) did not change much. The complexity of the MMs is not an issue nowadays, given the available computer power. Furthermore, the unlimited computer power enables new techniques for model extraction, and ANN is gradually replacing regression methods.

Note also the novelty of the iterative procedure adopted in Radojčić et al. (2017a, b) used to shape an incomplete database, eventually resulting in an MM for predicting resistance and dynamic trim of Series 62. To some extent this reverses the conventional sequential set of steps depicted in Fig. 2.1.

Development of MMs for R/Δ and τ with multiple-outputs (Radojčić et al. 2017b, and in addition for $S/\nabla^{2/3}$ and L_{WL}/L_P in Radojčić and Kalajdžić 2018) as opposed to those with a single-output, is a novel application of the ANN technique. Here too, the established reasoning that R/Δ and τ are only loosely related may be questioned, given that more than 92% of equation terms are common for the two different quantities. Note, however, that even a *loose connection* (in physical terms) between R/Δ and τ is often ignored, which is obviously wrong.

© The Author(s), under exclusive licence to Springer Nature Switzerland AG 2019 91
D. Radojčić, *Reflections on Power Prediction Modeling of Conventional High-Speed Craft*, SpringerBriefs in Applied Sciences and Technology, https://doi.org/10.1007/978-3-319-94899-7_7

New ANN based models, whether multiple- or single-output, clearly exhibit the double hump for dynamic trim. This was not possible with regression-based models as these were stiffer. Double hump in dynamic trim curve is important and may indicate dynamic instability.

When comparing MMs for resistance prediction with those for a propeller's hydrodynamic characteristics, note that the dependent and independent variables for the resistance prediction vary from model to model, depending on speed range and hull type. This is not the case with the propeller's variables which are essentially predetermined. For both the extraction tool used was regression and ANN.

MMs are routinely used to represent a propeller's hydrodynamic characteristics (K_T and K_Q) and the original experimental results (the dataset) are rarely given, nor are they needed. Knowing this, there is no reason why contemporary MMs for resistance and trim prediction (those that are based on a single systematic series) should not also replace the original databases. It appears that the MMs for resistance prediction are not yet fully accepted as authentic representatives of the datasets upon which they are based.

Given the experience with resistance modeling, it may be anticipated that ANN with multiple outputs may be applied for modeling propellers' hydrodynamic characteristics K_T and K_Q [as K_T and K_Q are linked through the open water efficiency coefficient—$\eta_o = (K_T/K_Q) \cdot (J/2\pi)$].

In summary, reliable MMs for prediction of resistance and propeller efficiency, although important, are not sufficient for reliable HSC power modeling. The interaction of hull, propulsor, and engine is the principal issue for appropriate calm water power predictions. Although there is room for improvements, the author believes that the integrated approach presented in Radojčić (1991) is fundamentally correct, and hence the subsequent MMs developed by him and his team were actually routines which can be incorporated in this or similar comprehensive power prediction routines.

Last but not least, the side-effect benefit of modeling is data-smoothing over continuous multidimensional surface, naturally when overfitting is avoided. From that perspective, MMs that are based on the systematic series may produce better results than those that stem directly from the original database. In some cases MMs were even able to identify erroneous datasets, and irregularities with the model experiments conducted years before these MMs were developed. This is contrary to the conventional wisdom that a MM, in general, cannot be better than the data upon which it is based.

References

Radojčić D (1991) An engineering approach to predicting the hydrodynamic performance of planing craft using computer techniques. RINA Trans 133

Radojčić D, Kalajdžić M (2018) Resistance and trim modeling of Naples hard chine systematic series. RINA Trans, Int J Small Craft Technol. https://doi.org/10.3940/rina.ijsct.2018.b1.211

Radojčić DV, Zgradić AB, Kalajdžić MD, Simić AP (2017a) Resistance and trim modeling of systematic planing hull Series 62 (with 12.5, 25 and 30 degrees deadrise angles) using artificial neural networks, Part 1: The database. J Ship Prod Des 33(3)

Radojčić DV, Kalajdžić MD, Zgradić AB, Simić AP (2017b) Resistance and trim modeling of systematic planing hull Series 62 (with 12.5, 25 and 30 degrees deadrise angles) using artificial neural networks, Part 2: Mathematical models. J Ship Prod Des 33(4)

Correction to: Reflections on Power Prediction Modeling of Conventional High-Speed Craft

Correction to:
D. Radojčić, *Reflections on Power Prediction Modeling of Conventional High-Speed Craft*, SpringerBriefs in Applied Sciences and Technology, https://doi.org/10.1007/978-3-319-94899-7

The original version of the book was inadvertently published without incorporating the following corrections:

In Chapter 1, Page 1, paragraph starting with the text: "The core of these work … required for design optimization" is repeated below. So the first instance should be deleted and the second one should be retained.

In Chapter 3, Tables 3.1, 3.2 and 3.3 need to be used in same formatting, either in portrait or in landscape.

In Chapter 4, Page 58, the symbol "v" should be replaced with "v" in three of the occurrences.

The correction chapters and the book have been updated with the changes.

The updated version of these chapters can be found at
https://doi.org/10.1007/978-3-319-94899-7_1
https://doi.org/10.1007/978-3-319-94899-7_3
https://doi.org/10.1007/978-3-319-94899-7_4

Printed in the United States
By Bookmasters